Palgrave Studies in Anthropology of Sustainability

Our series aims to bring together research on the social, behavioral, and cultural dimensions of sustainability: on local and global understandings of the concept and on lived practices around the world. It publishes studies which use ethnography to help us understand emerging ways of living, acting, and thinking sustainably. The books in this series also investigate and shed light on the political dynamics of resource governance and various scientific cultures of sustainability.

More information about this series at
http://www.springer.com/series/14648

Peter Bille Larsen • Dan Brockington
Editors

The Anthropology of Conservation NGOs

Rethinking the Boundaries

Editors
Peter Bille Larsen
Department of Anthropology
University of Lucerne
Luzern, Switzerland

Dan Brockington
Institute for International Development
University of Sheffield
Sheffield, South Yorkshire, UK

Palgrave Studies in Anthropology of Sustainability
ISBN 978-3-319-86894-3 ISBN 978-3-319-60579-1 (eBook)
DOI 10.1007/978-3-319-60579-1

Printed on acid-free paper

This Palgrave Macmillan imprint is published by Springer Nature
The registered company is Springer International Publishing AG
The registered company address is: Gewerbestrasse 11, 6330 Cham, Switzerland

Dedicated to the conservation practitioners,
our continuous conversations,
and their tireless efforts to make a difference.
—Peter Bille Larsen

For Tekla, Emily, and Rozie.
—Dan Brockington

Contents

List of Contributors

Peter Bille Larsen Department of Anthropology, University of Lucerne, Lucerne, Switzerland

Libby Blanchard Department of Geography, University of Cambridge, Cambridge, UK

Steve R. Brechin Department of Sociology, Rutgers, The State University of New Jersey, New Brunswick, NJ, USA

Dan Brockington Institute for International Development, University of Sheffield, Sheffield, South Yorkshire, UK

David Cleary Director of Agriculture, The Nature Conservancy, Arlington, VA, USA

Janet Fisher School of Geosciences, University of Edinburgh, Edinburgh, UK

George Holmes School of Earth and the Environment, University of Leeds, Leeds, UK

Ashish Kothari Kalpavriksh Environment Action Group, Pune, India

Siddhartha Krishnan Ashoka Trust for Research in Ecology and the Environment, Bangalore, India

Richard Ladle Institute of Biological Sciences and Health, Universidade Federal de Alagoas, Maceió, Brazil

Ken MacDonald Department of Geography and Center for Diaspora and Transnational Studies, University of Toronto, Toronto, ONM1C 1A4, Canada

Marianne Manuel Dakshin Foundation, Bangalore, India

Emmanuel Nuesiri University of Potsdam, Potsdam, Germany

Kent Redford Archipelago Consulting, Portland, ME, USA

Diane Russell Independent Consultant, Washington, DC, USA

Denis Ruysschaert CERTOP, University of Toulouse Jean Jaurès, Toulouse, France

Denis Salles Amenities et Dynamique Des Espaces Ruraux (ADBX), IRSTEA, Bordeaux, Cestas, France

Chris Sandbrook UNEP World Conservation Monitoring Centre, Cambridge, UK Department of Geography University of Cambridge, Cambridge, UK

Katherine Scholfield Synchronicity Earth, London, UK

Kartik Shanker Ashoka Trust for Research in Ecology and the Environment, Dakshin Foundation and Indian Institute for Science, Bangalore, India

Ed Tongson Sustainability Technologies, WWF Philippines, Quezon City, Philippines

Bhaskar Vira Department of Geography, University of Cambridge, Cambridge, UK

David Wilkie Conservation Measures and Communities, Wildlife Conservation Society, New York, NY, USA

List of Figures

List of Tables

CHAPTER 1

Introduction: Rethinking the Boundaries of Conservation NGOs

Peter Bille Larsen and Dan Brockington

Introduction

As debates rage on about changes required to build a different future for the planet, the role of conservation nongovernmental organisations (NGOs) as the global watchdogs of sustainability is increasingly prominent, but also questioned, in the public sphere. Vigorous debates about the role and effects of conservation NGOs call for independent analysis and debate about contemporary challenges and solutions. This book aims to showcase and challenge some of the latest engagements between critical social science and conservation NGOs. The authors have sought to do this partly because they believe it to be fundamentally important. Through such engagements it is possible to learn more about the consequences and politics of conservation policy, the way in which organisations function, and the interactions between various epistemologies and epistemic communities. This is a productive and

P.B. Larsen (✉)
Department of Anthropology, University of Lucerne, Lucerne, Switzerland

D. Brockington
Institute for International Development, University of Sheffield, Sheffield, South Yorkshire, UK

P.B. Larsen, D. Brockington (eds.), *The Anthropology of Conservation NGOs*, Palgrave Studies in Anthropology of Sustainability, DOI 10.1007/978-3-319-60579-1_1

insightful area for both researchers and practitioners. The chapters that follow showcase and debate some of the approaches that demonstrate these insights.

The authors also wanted to bring the chapters in this book together because the engagements can occasionally be frustrating. Too much of it is played out in contentious and adversarial ways that makes mutual learning and exchange difficult. The point is not that all the conflict is unwelcome, for some of these issues are best approached agonistically, where the difference is resolved through public debate, claiming space and airing of differences (Matulis and Moyer 2016). Consensus can be stifling and cloying. Still, however, it does not seem that the balance is right. Too many of the antagonisms on which Redford commented (2011 and reproduced in Chap. 9 of this book) remain. Therefore, there is also a commentary on, and an attempt to shift, the tone of these interactions. As will be clear from the commentaries (see particularly commentaries by Wilkie and Cleary), this book itself shows that there is still much work to be done.

Anthropological interest is not merely about whether NGOs make a difference but, paraphrasing Gregory Bateson, about understanding the difference that makes a difference (Bateson 1973). The real interest involves contributing to a more multifaceted understanding of NGOs, their forms of action and the contextual realities within which they operate. How is it possible to represent what conservation NGOs are and what they do if we acknowledge that they are dynamic and made up of webs of relations and networks rather than monolithic entities? Are anthropological and related critiques one step behind a dynamic reality, or one step ahead in terms of shedding light on NGO practice? Are conservation NGOs, in turn, ready for or resistant to 'informed criticism' (MacDonald 2003)? Are academics able to speak to the complex realities of conservation professionals and activists, and are the latter ready to explore and challenge basic assumptions and contentious politics? Where conservation NGOs look for success stories to describe achievements or give a positive spin, are social scientists, in turn, overemphasising the flipside of the coin? As Brosius noted:

> Anthropologists are seen to be fiddling while Rome burns. Furthermore, what anthropologists view as critiques derived from a particular set of theoretical premises, those in the conservation community view as criticisms, and this creates resentment. The fact that anthropologists, although prepared to critique, often fail to provide alternatives, only reinforces the perception that their criticisms are corrosive, irresponsible, and without validity. (2006: 684)

The discontent with critical literature and the difficulties of meaningful engagement persist a decade later. Neither the chapters, nor commentaries represented here shy away from strong positions, leading to some frustration in terms of nurturing fruitful exchanges. Can we, as Ashish Kothari (see Ashish Kothari (Chap. 12) of this book) and others, call for 'further shed[ding] stereotypes and be[ing] open to collaborations that can make conservation more effective and also more democratic and socially sensitive'? Notably, many of the commentaries reacting from the conservation NGO field, express more alignment with Redford's summary of social science contributions in Chap. 9 compared to other chapters. One reason, we suggest, concerns the challenge of representation, translation and contextualization in anthropology. Another, concerns the potential differences of perspective. What is certain is that there is a need for both more debate and multiple perspectives.

Nonetheless, first things first—this book is about conservation NGOs, but what do we mean by that? What constitutes a conservation NGO? The answer may appear straightforward, yet the ever-changing faces of conservation NGOs, histories of transformation from protest to advocacy, business or public service delivery point to significant differences, not merely subtle variations. The chapters herein challenge common assumptions about who and what conservation NGOs are and what they do. The sheer diversity of conservation NGOs both in terms of internal differences, underlying structures and evolving practices make them dynamic social entities. Compared to government hierarchies, NGOs' structures are more flexible, responsive to project funding and shifting dynamics. From Latin America to Asia, the NGO scene includes both affiliates of international organisations and homegrown institutions varying considerably in terms of political weight, constituencies and action forms (Miller 2007).

Many environmental organizations have shifted from initial positions of advocacy and confrontation to cooperation and interaction (Kraft 2001). One study of environmental justice organizations suggests a shift toward formalization, partnerships and networking (Perez et al. 2015). In Asia, for example, it has been suggested that policy influence on domestic environmental NGOs is limited compared to wider global processes (Frank et al. 2007); although, some national organizations, as in Indonesia, have been pivotal in influencing governmental environmental policies (Ruysschaert 2013). One of the challenges of this debate is to point to trends and patterns while capturing the diversity involved.

Conservation NGOs as Boundary Organizations

The idea of conservation NGOs as 'boundary organizations' offers one gaze to recast the debate. The concept has been used to describe organisations working at the boundaries between science and politics (Guston 2001; Carr and Wilkinson 2005). We suggest here a more expansive approach conceptualising boundary organisations as covering a broader set of boundary interactions, identities and relationships (O'Mahony and Bechky 2008). Organizations do not operate as self-contained entities in isolation, but evolve through boundary interaction with a variety of networks, multiple sectors and institutional contexts well beyond the science–policy interface. This entails roles of reshaping and defining the contours of conservation concerns, identities and constituencies as well as ways of framing and positioning themselves ideologically in relation to other actors. Thus, while conservation etymologically is about preserving and maintaining something, practice entails constant responses to and engagement with changing social, political and economic boundaries. Larsen (Chap. 2 of this book) notes how this creates friction between commonly held ideas of pure conservation action versus publicly contested forms of (inter)action.

Conservation NGOs inevitably entail interactions with a wide range of actors. Where business and government may appear as odd bedfellows with conservation on paper, today they are regular partners in more explicit terms. The question is not whether this takes place or not, but rather understanding the implications of these entanglements, trans-boundary transactions and the choices behind them. Ranging from explicit strategic engagement to tacit involvement, boundary interaction from conflict and contestation to cooperation is shaping the nature and outcomes of conservation action.

It is increasingly difficult to maintain divides between domestic and/or international, civil society (i.e., the state) or conservation and development. Faced with daunting climate change and biodiversity loss challenges, boundaries are continuously challenged to explore new frontiers of action. The ensuing questions of benefits and costs of this new set of interactions are frequent in the public sphere between organisations working with, and those challenging, the mainstream (McDonald 2016). What from one perspective may represent win–win gains, of shifting practice through alliance building and conversation, is from another perspective seen as giving in to the status quo.

Redefined roles, in relationship to social movements and grassroots organizations, business partnerships and state politics, are part and parcel of an NGO's life. Specific sectoral negotiations, or campaigns, often reveal changing alliances, diverse positionalities and intrasectoral divergences among global NGOs (Pallas 2013). Where conservation NGOs may engage in boundary maintenance to communicate and single out their core values and distinctive roles in society, they may equally challenge boundaries and venture into new forms of action. Engagements in social justice issues, for example, are not uniform, and complexities are rarely evident in the policy statements and self-representations of conservation action even within one NGO. Indigenous representatives may appear as board members in one context, be offered central roles in the development of policy standards, yet remain outside decision-making contexts in others.

Interrogating the *non*governmental of conservation, in this respect, concerns one important boundary. For one, certain forms of 'NGO' action today are often far more governmental than the name and history suggests. Red tape, permits and control, but also collaborative funding arrangements, capacity-building and long-term partnerships with state agencies are part and parcel of conservation work. Blurred boundaries are the rule rather than the exception in the (non)governmental sphere. Relationships with the state remain fraught with complexity between complicity, outright contestation of some ministries and decisions, while delivering services to and building capacity of others. Indeed, state machineries are equally complex and contradictory, rendering sensitivity to networks and 'transboundary' activity a critical feature of ethnographic attention to state-related NGO action.

Several papers cross-examine the implications of NGOs moving closer to, while speaking to and struggling against power. Conservation NGOs also display power and influence, and therefore merit analytical attention. The range includes conservation NGOs speaking with power and authority in local settings to internal power struggles or power implications of transboundary alliances with business and the state.[1]

In 2016, for example, Survival International launched an official complaint against the World Wildlife Fund (WWF), under the Organisation for Economic Cooperation and Development (OECD) Guidelines for Multinational Enterprises, for having contributed to violence against the Baka in Cameroon through support of protected area creation and ecoguards targeting Baka in Southeastern Cameroon. The allegations suggested the centrality of NGO support in maintaining—and failing to secure the respect of human rights—in a highly unjust system, nurturing displacement and even violence (Survival International 2016). This situation was not as unusual as it might appear.

Over the last couple of decades, human rights and local NGOs have denounced the impacts of conservation NGOs (Colchester 2003; Pyhälä et al. 2016). The relationship between conservation and human rights remains fraught and uncertain (Winer et al. 2007). Some NGOs' call for and support to militarized enforcement to fight the wildlife crisis is increasingly present in parts of the world (Duffy 2014), resulting in 'green militarization', which is understood as the use of military and paramilitary technologies in the pursuit of conservation (Lunstrum 2014). Still, the analytical point is not one of displaying NGO power alone, but of recognising the diversity of the power relationships at stake.

Power deficiencies, weakness and short-lived windows of opportunity of NGOs are far more frequent than actual influence. Furthermore, shifting alliances over time are not captured by the somewhat simplistic opposition between international NGO power trumping local action. Many NGOs today engage in alliances and collaborative efforts with indigenous people, local communities and their organizations. This book therefore insists on understanding NGOs as diverse and constituted both by varied webs of relations, responses to the contexts within which they operate, and portfolios of activities.

Boundaries of Action

Whereas NGOs generally were recognized as 'non-institutional' agents of both democratisation and development in the 1980s (Lewis 2016), the post-Rio (1992) prominence of conservation NGOs involved a window of privileged agency of sustainability and incremental institutionalisation both nationally and internationally. Conservation NGOs have contributed both to addressing and defining the sustainability frontiers and the normative frameworks of our times. Where it is common to distinguish between mainstream and radical organizations, NGOs are rarely confined to one approach, nor one form of engagement. Campaigns, business partnerships and legal action exist side by side without being flagged as such, and often evolve over time. Protest movements may become mainstream partnership organisations engaging with state power, just as popular support may drive mainstream organisations into campaign mode. The NGOs' actions and effects are not easily captured in standard typologies.

Amid the diversity of responses visible in the plethora of conservation NGOs' actions, some interesting contrasts are evident not the least being international meetings. Consider the differences of discourse between the 5th International Conference on Degrowth for Ecological Sustainability and Social Equity, 'Walking the Meaningful Great Transformations' (Budapest, September 2016), and the World Conservation Congress, 'Planet at the Crossroads', in Honolulu attended by some 10,000 participants in the same year (see also MacDonald, Chap. 4 of this book). The Budapest conference spoke of 'Degrowth in Practice' under the motto 'slow down and think'.[2] Panel topics included: Connecting the Dots of Degrowth, the Growth Economy and Challenges for a Social Ecological Transformation and From Capitalist Accumulation to a Solidarity Economy.

The Hawai'i 'Commitments' in turn spoke of diverse voices finding common ground 'in a spirit of partnership and collaboration', a 'culture of conservation' and building constituencies aiming for '[e]conomic and legal systems … that reward communities and companies for actions and investments that protect and restore nature'. This conference called for more public and private investments in conservation and a 'collaborative approach, including government, civil society and the private sector' as 'essential for success'.[3]

A certain form of sanitized politics prevails and currents of partnership, dialogue and reform language clearly prevail over radical change among international conservation NGOs. *Dialogue* is prioritized as a value and engagement strategy. Ranging from staged 'high-level' encounters to institutional dialogue and roundtable mechanisms, how do impacts differ compared to more radical questioning and positions characterized by protests or confrontational media campaigns?

The NGOs and their operations are not simply signs of our times, but involve the pursuit of distinct ideas (Blanchard et al., Chap. 6 of this book) and distinct forms of agency and engagement (Ruyschaert and Salles, Chap. 5 of this book), chasing emerging opportunities through private financing, social media attention and political change. They are not only users but equally creators and sellers of conservation tools and approaches. In the often fiercely competitive marketplace of conservation finance, singularity and added-value are part and parcel of winning competitive bids and a seat at the table. Such conservation entrepreneurialism and creative thinking obviously entail risks and opportunities. A number of the authors (Ruyschaert and Salles, and Nuisiri, Chaps. 5 and 8 of this book) explore and challenge as to whether conservation NGOs achieve conservation and wider development goals with their current portfolios. Critical assessments of effectiveness and equity impacts, however, should not overshadow that progress may be achieved through various forms of engagement (e.g., see Cleary, Chap. 11 of this book), nor the potential lost opportunities where more radical positions may deliver significant long-term results.

The Contents of This Book

The idea behind this book came out of a workshop, entitled 'Anthropology of Conservation NGOs', held in Chicago in December 2013. The themes and perspectives struck a chord with a growing call for further engagement and bridge-building between critical research and conservation NGOs. Additional contributions were brought together and commentaries from practitioners and observers of the NGO field were solicited to trigger and stimulate further debate.

One of the central themes of the chapters here is the changing relationship of conservation organisations with the market-based conservation and neoliberal capitalism more generally. Libby Blanchard, Chris Sandbrook, Janet Fisher and Bhaskar Vira look at the variety of attitudes toward market-based instruments using the Q methodology among conservationists at the International Congress for Conservation Biology (in New Zealand in 2011) and among conservation organisations clustered in Cambridge, UK (in 2013). They find a divergent set of attitudes toward the use of market-based instruments in conservation. At both sites there was a clear group of market enthusiasts, who embraced the opportunities that markets provide. As one of the respondents put it:

> [W]e used to be combative and confrontational, presenting to the rest of the world capitalism as the cause of the decline in biodiversity. Now we are moving into a much more mature frame of mind that says collaboration. Let's try to solve these problems together. Let's take what money, wealth, and capitalism can do at face value and help it do the right thing to make the world a better place.

The strength of the common viewpoints in New Zealand and Cambridge over a two-year period suggested a single epistemic community forming around embracing and supporting market-based initiatives in conservation. There were some indications that the move to embrace markets is being led from the top by leaders in the respective organisations that took part. Nevertheless, there also were a series of viewpoints that expressed more scepticism of, if not opposition to, market enthusiasm. These were, however, more fragmented, and expressed in a different way across the two study sites the researchers dealt with.

Similarly, George Holmes, in Chile, details the surprisingly opportunistic and contingent engagement of conservation organisations in setting up new protected areas on private land. These are happenstance alliances not driven by large-scale, long-term planning, but simply because new opportunities to set aside large areas of land emerged. The growth of protected areas here came about, ironically, because of legislation designed to make it easier for companies to take over land for development. They (the protected areas) are flourishing in part because capital investments are not. These are marginal lands that can be set aside for conservation because commercial interest in them has waned. This strikes a chord with a broader argument about the significance of context to fully grasp the action realm of conservation NGOs.

Ken MacDonald's chapter tackles one of the most vibrant aspects of the marriage between mainstream capitalism and mainstream conservation head on, examining the performance and institutionalisation of the private sector with conservation organisations at the World Conservation Congress (WCC) in 2008. Through exploring the political message, spectacle, orchestration and organisation of the conference he found that:

> The 2008 WCC provided a notable window into the consolidation of such relationships, perspectives, and processes and their role in shaping the new organisational order of IUCN; one which situates markets, business, and private sector actors firmly at the core of the Secretariat, if not the membership, of IUCN.

Engagements with capitalism also frame the work of Denis Ruysschaert and Dennis Salles on the NGOs engaged in the Roundtable on Sustainable Palm Oil (RSPO), a private multi-stakeholder initiative to produce 'sustainable' oil palm. Nonetheless, their case serves to illustrate a further point—that is, the work of conservation NGOs is best understood by exploring the work of collectives of organisations and by exploring the fault lines and continuities across them. This chapter examines the paradox of NGO engagement with palm oil and tropical deforestation with a focus on Southeast Asia. There, both deforestation for palm oil and engagement in the RSPO are increasing. What drives NGO involvement and these apparent failures in the Roundtable's ultimate goal? Based on more than five years of engagement with the RSPO, the authors divide NGOs into those who collaborate, those who oppose, and those who are opportunists and chose either collaborative or oppositional stances depending on the needs in specific areas. The last group, the sceptics, are concerned more with community land rights than collaboration. A key limitation was that NGOs in these diverse strategic camps were separated in practise. They could not collaborate across the strategies.

Dan Brockington, Katherine Scholfield and Richard Ladle take a sectorial approach, considering the work of NGOs as united by their geography—in this case, their work in sub-Saharan Africa. Using previously published work, which describes the distribution of activities, they show that several thought experiments and various insights into their work and performance become possible with this perspective. Indeed, Scholfield's observation, from her study of one conservation NGO network in Rwanda, was that 'NGO activities made little sense when viewed as the work of single organisations'. This approach also can indicate new possibilities of collaboration and make it possible to ask which synergies might be possible if forces were combined. Their challenging observation is that entirely new scales of collaboration could result in transformative support for protected areas. Current estimates suggest that it would cost a little over USD60 million (in 2006) to fully protect more than 90% of the more strictly protected areas (IUCN categories I–IV). This is without further state subsidy. Expenditures in 2006 by conservation NGOs was probably more than three times that figure.

Emmanuel Nuesiri's concern is with the work of the NGO sector on the local and national scale. Building on work by Ribot and others, he observes that conservation NGOs have been used effectively to circumvent local process and politics with respect to the implementation of the UN Programme for Reducing Emissions from Deforestation and Forest Degradation (REDD+) in the Cross River State in Nigeria. Their presence made it possible for forestry groups that were implementing REDD+ to work without going through elected local government representatives. At the same time these NGOs were not well equipped to challenge some aspects of the REDD+.

Bridging the gaps between critical social scientists working on conservation, and many in the conservation community remains difficult. This is clear in the work of Blanchard and colleagues, who found that there seemed to be little connectivity between critical social science scholars' views about market-based instruments and the market sceptics in conservation communities. Peter Bille Larsen in his Chap. 2 emphasizes how much both public and scientific discourse are framed easily in simplistic narratives, ill-adapted to capture the complexity and changing practices of conservation NGOs. Even though conservation NGOs in part thrive on narratives of fighting for the public environmental good, counternarratives about NGOs growing big, 'ugly' and business-minded are common in both social science and public discourse. Through a case study of the shifting roles and practices of NGOs in the Peruvian Amazon, Larsen calls for a critical middle ground of analysis that will be able to capture problematic spaces as well as alternative institutional forms and practice worthy of anthropological exploration. Such middle grounds, however, are perhaps better thought of as turbulent seas.

Kent Redford's Chap. 9 (and David Wilkie's commentary) offer salutary and critical observations as to the way that this engagement is undertaken. Redford's essay, which has been reproduced in this book, provides a short list of the ways in which social science has improved conservation practice, and a lengthier discussion of the problems with social science engagement. Social science analysis does not appreciate sufficiently the diversity of conservation practice, thinking, strategy or instruments. Redford remains hopeful that engagement between diverse sides can be constructive, and he calls for continued engagement with *informed* (our emphasis) social science.

The final section of the book is a collection of independent commentaries by thinkers and practitioners who are active in the conservation NGO field on writings we gathered and have just summarised. Solicited in the spirit of nurturing further debate, the diversity of responses is encouraging as well as indicative of vigorous debates ahead. On the one hand, there is general agreement about the significance of social sciences within and about the role of conservation NGOs. Ed Tongson, for example, underscores how stakeholder analysis, participatory research and approaches to equitable sharing of costs and benefits informed the toolkit employed in marine conservation. Diane Russel, in turn, offers a succinct perspective on the importance of institutional factors in shaping NGO action. Kartik Shanker, Siddhartha Krishnan and Marianne Manuel, in a point of view from India, challenge dichotomies between local and international NGOs, noting similitudes but also variation and complexity.

On the other hand, several commentaries express uneasiness with the theoretical premises and analytical value of critical approaches. David Cleary challenges analysis as being rigidly grounded in a political economy framework and ideology that disregards actual achievements. David Wilkie equally suggests that language and an overemphasis on neoliberalism from afar hinders effective dialogue with conservation practitioners. Steve Brechin, in contrast, calls for renewed debate about NGO ethics, reminding readers of attempts to silence a critical essay more than a decade ago (Chapin 2004).

Debates around language, theory and perspective may at first glance appear unusually heated and emotional, yet are also signs of the profound engagement found in both the social sciences and conservation NGOs. These are not matters taken lightly either at a personal or an organizational level. The discussions offer a refreshing basis for exchange and alternatives to the polished nature of dialogues predominating contemporary conservation forums.

Concluding Remarks

The shifting boundaries of conservation NGO identities and actions offer an important terrain of study as well as fields of engagement and exchange. Yet, at the end of the day, do critical approaches and further anthropological analyses matter? What difference, if any, do they make? Both the findings and discussions offered here testify to the fertile grounds.

Kothari, in his commentary, calls for 'rethinking epistemologies based on much closer, collaborative work with indigenous peoples and other traditional local communities'. This challenge arguably goes for both the social sciences, such as anthropology, and conservation NGOs. Although specific anthropological studies or conservation projects may come under attack for imposing their knowledge forms, this should not draw attention away from the fact that overall global trends of sidelined science, immobilized NGOs and climate change responses are being hollowed out. In this sense, social sciences share a fundamental epistemological challenge in terms of the various ways of knowing and, ultimately, stewarding the world. Central to this endeavour are the multiple ways in which human environment relationships, practices and values are conceptualized—a central theme in both environmental anthropologies and political ecology at large.

Given that NGOs are a critical piece in the larger puzzle of biodiversity conservation, the ability to interrogate and speak to the concepts, approaches and partnerships undertaken, and how they relate to social and ecological processes, remains important. Such attention needs to be directed not only at NGOs themselves, but also through boundary engagement to shed light on wider trends of sustainability politics, economic organization and social practise. This entails shifting from individual projects and activities toward contextually informed organizational histories and sectoral perspectives. In particular, tracing the craft and long-standing moral guardians of sustainability needs to be seen in the context of nation-state policy and global inequalities. The NGOs seek to occupy a powerful institutional space, yet cultural brokers and boundary actors are equally vulnerable to governmental control and the perils of unstable finance.

Understanding the effects and the effectiveness of conservation interventions continues to trail most other policy fields (Ferraro and Pattanayak 2006). Responding to such dynamics, not the least being the mediating role of conservation NGOs in global and local responses to conservation challenges, is key to current and future sustainability interventions. Consistent calls for strengthening the evaluation of conservation measures raises the need for applied social science, as well as critical approaches that reach beyond immediate output and outcome indicators. A given initiative may be judged relevant or effectively implemented, but still not necessarily capture shifting institutional arrangements, political trends and future viability. Given the magnitude and multiple scales of socioenvironmental challenges, the ways in which problems are framed and solutions play out in practice in diverse social, political and cultural contexts should be at the heart of research.

Sherry Ortner recently asked: 'How can we be both realistic about the ugly realities of the world today and hopeful about the possibilities of changing them?' She was describing how 'dark anthropology' since the 1980s has tended to focus on 'the harsh dimensions of social life (power, domination, inequality, and oppression)' (Ortner 2016). The question is of more obvious relevance than ever if individuals take the sustainability challenges of our times seriously. Whereas the necessity of 'dark' analysis remains, there is certainly room for additional empirical and theoretical ventures into the world of conservation NGOs.

Anthropology from this perspective need not be about jumping on the bandwagon of imagining futures and a positive solution spin, but first and foremost of decrypting practice in real social terms. Applied conservation anthropologies have a long history of engagement through, for example, ethnobotany and problem and stakeholder analysis; however, equally so through employing critical social science and political ecology to contextualize the reach of conservation NGOs from a dynamic, relational and processual perspective. Dramatic environmental transformations prompt not only attention to changing state policies and multilateralism, but equally so to the efforts and context of conservation NGOs' actions.

Notes

1. Note that Sandbrook's blog on this topic is available at: https://thinkinglikeahuman.com/2016/09/22/weak-yet-strong-the-uneven-power-relations-of-conservation/
2. http://budapest.degrowth.org/. Accessed 18 December 2016.
3. https://portals.iucn.org/congress/sites/congress/files/EN%20Navigating%20Island%20Earth%20-%20Hawaii%20Commitments_FINAL.PDF. Accessed 18 December 2016.

References

Bateson, G. 1973. *Steps to an Ecology of Mind: Collected Essays in Anthropology, Psychiatry, Evolution and Epistemology*. London/Toronto: Paladin Granada Publishing.

Brosius, P. 2006. Common Ground Between Anthropology and Conservation Biology. *Conservation Biology* 20 (3): 683–685.

Carr, A., and R. Wilkinson. 2005. Beyond Participation: Boundary Organizations as a New Space for Farmers and Scientists to Interact. *Society & Natural Resources* 18 (3): 255–265.

Chapin, M. 2004. A Challenge to Conservationists. *World Watch Magazine*, November/December 2004.

Colchester, M. 2003. *Salvaging Nature: Indigenous Peoples, Protected Areas and Biodiversity Conservation*. Montevideo: World Rainforest Movement and Forest Peoples Programme.

Duffy, R. 2014. Waging a War to Save Biodiversity: The Rise of Militarized Conservation. *International Affairs* 20 (4): 819–834.

Ferraro, P.J., and S.K. Pattanayak. 2006. Money for Nothing? A Call for Empirical Evaluation of Biodiversity Conservation Investments. *PLoS Biol* 4 (4): e105. doi:10.1371/journal.pbio.0040105.

Frank, D.J., et al. 2007. World Society, NGOs and Environmental Policy Reform in Asia. *International Journal of Comparative Sociology* 48 (4–5): 275–295.

Guston, D.H. 2001. Boundary Organizations in Environmental Policy and Science: An Introduction. *Science, Technology, & Human Values* 26 (4): 399–408.

Kraft, M.E. 2001. Influence of American NGOs on Environmental Decisions and Policies: Evolution over Three Decades. In *The Role of Environmental NGOs: Russian Challenges, American Lessons: Proceedings of a Workshop*, 145–160. Washington, DC: National Academy Press.

Lewis, D. 2016. Anthropologists' Encounters with NGOs: Critique, Collaboration, and Conflict. In *Cultures of Doing Good: Anthropologists and NGOs*, ed. A. Lashaw, C. Vannier, and S. Sampson. Tuscaloosa: University of Alabama Press.

Lunstrum, E. 2014. Green Militarization: Anti-Poaching Efforts and the Spatial Contours of Kruger National Park. *Annals of the Association of American Geographers* 104 (4): 816–832.

MacDonald, K. 2003. *IUCN: A History of Constraint.* Text of an Address Given to the Permanent Workshop of the Centre for Philosophy of Law Higher Institute for Philosophy of the Catholic University of Louvain (UCL), Louvain-la-neuve.

Matulis, B.S., and J.R. Moyer. 2016. Beyond Inclusive Conservation: The Value of Pluralism, the Need for Agonism, and the Case for Social Instrumentalism. *Conservation Letters*: n/a-n/a.

McDonald, M. 2016. Bourdieu, Environmental NGOs, and Australian Climate Politics. *Environmental Politics* 26 (6): 1058–1078.

Miller, S. 2007. *An Environmental History of Latin America.* Cambridge: Cambridge University Press.

O'Mahony, S., and B.A. Bechky. 2008. Boundary Organizations: Enabling Collaboration Among Unexpected Allies. *Administrative Science Quarterly* 53 (3): 422–459.

Ortner, S. 2016. Dark Anthropology and Its Others: Theory Since the Eighties. *Journal of Ethnographic Theory* 6 (1): 47–73.

Pallas, C. 2013. *Transnational Civil Society and the World Bank: Investigating Civil Society's Potential to Democratize Global Governance*. New York: Palgrave Macmillan.

Perez, A.C., et al. 2015. Evolution of the Environmental Justice Movement: Activism, Formalization and Differentiation. *Environmental Research Letters* 10 (10): 105002.

Pyhälä, A., et al. 2016. *Protected Areas in the Congo Basin: Failing Both People and Biodiversity?* London: Rainforest Foundation UK.

Ruysschaert D. 2013. Le rôle des organisations de conservation dans la construction et la mise en oeuvre de l'agenda international de conservation d'espèces emblématiques: le cas des orangs-outans de Sumatra. PhD Thesis, University of Toulouse. Available through: https://tel.archives-ouvertes.fr/tel-00951940

Winer, N., et al. 2007. Conservation, Protected Areas and Humanitarian Practice. *Policy Matters* 15: 232–240.

CHAPTER 2

The Good, the Ugly and the 'Dirty Harry's of Conservation: Rethinking the Anthropology of Conservation NGOs

Peter Bille Larsen

Introduction[1]

'Are you in?' I was asked this in the email header from the World Wide Fund for Nature (WWF),[2] a big conservation organisation, with which I have collaborated over the years. 'Do you care about a clean, healthy future for people and the planet?', the mass mailing continued. The following section, 'Our Pledge', noted: '[W]e believe our future should be powered by nature' and emphasised the need for 'investments in clean and renewable energy'. It went on: 'We choose to invest in solutions, not in problems'. The email's message ended with: 'Click "yes" to sign our plea: seize your power'. Such power could either be seized through Facebook, Twitter or Google+, revealing the social media version of 'signing up' to 'good' solutions spearheaded by nongovernmental organisations (NGOs).

Conservation NGOs, dedicated to biodiversity at large, today form a natural part of the institutional landscape and public space. The NGOs'

P.B. Larsen (✉)
Department of Anthropology, University of Lucerne, Lucerne, Switzerland

P.B. Larsen, D. Brockington (eds.), *The Anthropology of Conservation NGOs*, Palgrave Studies in Anthropology of Sustainability, DOI 10.1007/978-3-319-60579-1_2

influence and presence grew exponentially in the years following the 1992 Earth Summit, leading to the expansion of field offices of Northern NGOs as well as the mushrooming of Southern conservation NGOs (Levine 2002). They have been particularly influential in shaping public opinion and policy in the North (Corson 2010) as well as in influencing policy terrains in the global South that harbour the highest concentrations of biodiversity.

Anthropologists have become increasingly active both in terms of working within and studying the work of conservation NGOs. First, conservation NGO projects are increasingly present in 'ethnographic' field settings because of the explosion of conservation initiatives across the globe. This has led to many projects hiring practising anthropologists. Second, conservation presence has led to tensions and dynamics with indigenous people and local communities, triggering various forms of anthropological critique. Third, a growing body of analysis has increasingly taken up conservation NGOs as an object of study in their own right. Anthropologists have portrayed local perspectives in which global narratives prevail. They have undertaken global event ethnographies (Brosius and Campbell 2010; Corson and MacDonald 2012), site-specific analysis (West 2006) as well as comparative work (Brockington and Scholfield 2010). The discipline is also at the forefront of 'elucidating institutional developments and the forms of environmental surveillance and intervention it promotes' (Brosius 1999: 50).

Nevertheless, rather than resulting in a concerted anthropological conservation agenda, such engagement is pointing to a number of contradictions. Not only has the biodiversity crisis deepened during the same period that NGO activity has mushroomed, but also the very solutions conservation organisations propose are questioned and, according to some, are even aggravating the problem (Igoe and Brockington 2007). Critical voices point to mainstream organisations, particularly big international nongovernmental organisations (BINGOs) that are dedicated to anodyne advocacy rather than activism. They observe technically framed solutions and compromise replacing politics, corporate partnerships substituting critique, narrow environmental policy and single issues predominating over broad-scale sustainability politics (Barker 2010; Chapin 2004; Holmes 2011; Levine 2002; MacDonald 2010). Furthermore, there have been attacks in the global South against international conservation NGOs, which are perceived as foreign enterprises that undermine rather than support national civil society organisations, whether through co-opting leadership or draining resources intended for local conservation work. Where problem analysis has long been part of the conservation toolbox and a source of NGO authority, critical voices insist that conservation NGOs are the problem.

The title of this chapter derives from the polemic nature of the debate. I suggest that many debates can be understood through the lenses of three master narratives that frame the significance of conservation NGOs' activity. Such narratives, respectively, position NGOs as (1) doing good, (2) turning ugly or (3) acting pragmatically through what this author labels 'Dirty Harry' approaches. I consider these as master narratives given their meta-discursive role in delineating and defining (Bamberg 2005) how conservation NGOs are perceived and understood.

The first involves the 'good conservationist', a master narrative often apparent in foundation documents, public profiles and programmatic statements of conservation NGOs. 'Doing good', I contend, is an essential part of conservation NGO activity and legitimacy production where organisational identities and activities are framed as a matter of moral duty and grassroots intervention.

The second master narrative, 'the ugly conservationist', has become increasingly apparent in the last decade, reflecting the blurred boundaries between 'good' conservation and 'bad' states and corporations. This counternarrative stresses how conservation NGOs have 'turned' big and ugly, distanced themselves from local constituencies and sided with power.

The third master narrative, the 'Dirty Harry's of conservation emphasise pragmatic values and realism in response to the earlier critique. Partnerships and engagement with the 'bad' state and private sector are not considered problematic per se, but necessary to secure real-life change.

At a time when engagements, both within NGOs and critical analysis from the outside have blossomed, how do we address this narrative complexity? As a starting point, this chapter argues that such meta-narratives leave little analytical space to capture the complexity at stake. First, there are obvious risks with taking the 'doing good' discourse for granted by performing complicit analysis without fundamentally interrogating the institutional challenges and constraints at stake. Second, even though the 'ugly conservationist' critique offers a much-needed reality check of conservation NGOs, this arguably leads to an analytical impasse that is caught up in dichotomies of the 'good' intentions and an 'ugly' present. As such, they leave social science inadequately equipped to capture the shifting realities, roles and practises of actual NGO work. Third, 'Dirty

Harry' narratives, although illustrative of evolving power dynamics and institutional strategy, easily displace or render critical analysis 'disobedient' (Igoe et al. 2010).

The resulting *malaise* suggests a need to revisit the role of anthropological analysis of conservation NGOs. This section is divided into three major parts. The first part debates master narratives and claims in the literature of conservation NGOs, respectively, doing good, turning ugly and operating as 'Dirty Harry's. Subsequently, the second part debates the analytical implications and limitations of this critique through a case study from the Peruvian Amazon. The third part offers a synthesis of the dispute that argues for the need to rethink anthropological counternarratives from a less dichotomous perspective.

Methodology

This analysis is informed by a reflexive exercise based on long-term involvement with(in) conservation NGOs as a practitioner of what might be labelled 'conservation anthropology', as well as an observer of them as part of a broader research into environmental governance based on ethnographic fieldwork in Peru. Assessments of specific conservation NGOs frequently reveal how realities are often more complex and diverse than what appears in critical narratives. Whereas the latter may emphasise situations of top-down design, dispossession and social exclusion, there are just as many counter examples of NGO actors establishing new forms of alliances, breaking new ground and advancing social agendas in instances where government inertia prevails. Wholesale dismissals of monolithic NGOs do not capture such complexity, prompting the need for rethinking anthropological analysis from the outside, together with constructive engagement from inside.

To effectively address and decipher the nature of NGO critiques and to compare them with the complexity of field realities, this research combines a literature review with an extended case study method. The literature review was used to identify common narratives and explanatory arguments appearing both in the moral justification of NGO action and to capture the discourses of its critics. This involved reviewing literature produced by conservation NGOs along with public discourse and academic analysis, where critiques have emerged within the last decade. The second methodological axis involved ethnographic methods, which have proven to be particularly productive in terms of capturing the slippery nature of NGOs as an object of study. The extended case study is based on ethnographic fieldwork stays in the Peruvian high jungle between 2007 and 2013 (Larsen 2015). For a long time, the area has been an established region for conservation efforts, offering a pertinent case study to explore evolving roles, practises and perceptions of NGO activity. Field data is based on participant observation, document analysis and semistructured interviews with NGO and local representatives.

The Good Conservationist and the Bad 'Other'

A central master narrative among conservation NGOs couples public environmental good with wider narratives of biological crisis (Escobar 1998: 56). A useful example of such narratives of crisis and global action is the so-called 'Morges Manifesto'—the founding document leading to the establishment of the World Wildlife Fund (WWF); it has since been renamed the World Wide Fund for Nature. Its title, '*We Must Save the World's Wildlife: An International Declaration*', speaks for itself (Morges Manifesto 1961). As the Manifesto stated, using war metaphors, there were 'skilful and devoted men and admirable organisations ... battling at this moment on many fronts'. It was an 'emergency' with a 'vast number of fine and harmless wild creatures losing their lives, or their homes, in an orgy of thoughtless and needless destruction'.

In language resembling a humanitarian rescue mission, the 'wildlife emergency' stemmed from 'ignorance, greed, and folly'. Animals were being 'shot or trapped out of existence ... drowned by new dams, poisoned by toxic chemicals, killed by poachers for game, or butchered in the course of political upheavals'. In response, the World Wildlife Fund was set up to offer the public 'an effective simple means' to save the world's wildlife; it was to be a new organisation with 'easy channels for all who want to help' and 'raise massive resources for the cause and distribute resources where they are most needed'. Advancing civilisation had to be stopped or as the title of 1969 album with the Bee Gees, the Beatles and others singing for the WWF went: 'No One's Gonna Change Our World'. On the album, the Beatles introduced 'Across the Universe'. Cilla Black sang 'What the world needs now is love'. Rolf Harris sang about the 'Cuddly old Koala', Lulu interpreted 'I'm the Tiger' and Bruce Forsyth presented 'When I see an Elephant Fly'. Conservation NGOs were the moral guardians of the good in the 'senseless orgy' supported by the latest tunes of public culture. This *Zeitgeist* remains important to this day in the nature of conservation communication strategies.

Conservation NGOs are in many foundation narratives about the good against the 'bad other' that undermines the public environmental good. Now, I may be criticised for being unfair, even incorrect, in exemplifying the 'good' of NGOs with the emergency language from the late 1960s. The objectives and approaches of NGOs have evolved, it may rightfully be contended. Contemporary versions of the 'good conservationist' often are found no longer in the alarmist moral condemnation of civilisation and species loss, but increasingly in the positive techno-solution realm of sustainability guidance (Larsen 2013). In effect, conservation NGOs are in a constant process of (1) reinventing themselves, (2) redefining mission statements and (3) fine-tuning strategies into new forms of action. This author's purpose is, not here at least, to enter the particular battlefields of what constitutes a good or bad conservation strategy. I argue, however, that it is essential to recognise the importance of the 'good conservationist' narrative as a moral constitution and an unwritten social contract between conservation NGOs and broader society. Conservation NGOs, in this meta-narrative, represent the good conservation cause against 'bad reasoning'. It, as such, constitutes an existential reasoning based on moral grounds and values, on the one side, and technical competence and science on the other. This link was best articulated by the International Union for Conservation of Nature (IUCN) Secretary General Duncan Poore in 1977:

> The Union is concerned with values more, I would say, even more than with science. For science should be the servant, not the master of mankind. Our strategy must be firmly based in realism, but it must move ahead with vision. We should be the architects of guided change (call it development if you will)—guided change in the direction of increasing the well-being of humankind—not only the standard of living but the good life, but (and the 'but' is all important) in such a way that the potential of the biosphere to support this good life is not diminished. (Holdgate 1999: 137)

Central to this narrative is not merely agency outside the realms of the state, but also more fundamentally the moral high grounds of technically sound agency above the temptations of *real politik* and short-termism. Corson (2010: 578), for example, speaks of conservation organisations capitalising through idealised visions of themselves as representatives of civil society countering private forces.

'Good conservationist' narratives are no longer only about conviction and identity, but they are tied to carefully designed branding operations

(Rodríguez et al. 2007) and 'the spectacle of capitalist conservation' (Brockington and Duffy 2010: 473). Conservation NGOs have become 'super brands' with high levels of 'consumer trust' (Laidler-Kylander et al. 2007) leveraged by both the NGOs themselves and interested third parties, through partnerships and alliances by way of spectacular accumulation (Sachedina et al. 2010). 'If you have the will, we will show you the way', conservation NGOs tell the world through carefully orchestrated and competitive communication campaigns.

Such narrative positions justified the spectacular growth of sustainable development funding to conservation NGO work after the 1992 Earth Summit. Nongovermental organisations could deliver where states were absent, offering more efficient and competent solutions. Just as development NGOs experienced their heyday as alternatives to the failure and pitfalls of conventional development schemes, conservation NGOs initially thrived on the conservation boom of the 1990s. Nonetheless, where supporting conservation NGOs, like voting in a democracy, was a natural reflex of concerned citizens, arguably the picture has now changed somewhat.

The Ugly Conservationist

An increasingly common narrative portrays NGOs, no longer only as small, beautiful and 'doing good', but as turning big, 'ugly' and transnational. Public debate and a growing body of literature have within the last decade thrown into question the mandates, roles and effects of Northern NGOs. This critique has been launched particularly against the BINGOs. Such organisations, critics voice, do not thrive only on a moral mission and merit against all odds, but they increasingly rely on a capitalistic expansion of activities, public finance entanglements, and flawed corporate partnership projects that threaten what they set out to protect in the first place. Descriptions, mainly of BINGOs, have moved from portraying an idealist independent activity toward describing professionalised, managerial and internationally financed institutions. This has led to a shattering critique that erodes the moral premises of the 'good conservationist' narrative.

Why this change? The shift partially reflects a new empirical reality including the spectacular growth of conservation finance. Annual conservation funding is now estimated to be in the range of USD19.8 billion (Waldron et al. 2013). Today NGOs are not only among the biggest investors in conservation (Khare and Bray 2004), but also equally central in the management and spending of these resources. Contemporary NGO action cannot be understood outside this political economy. Challenging the idea that NGOs signal a strong civil society, wider analysis has pointed to the material pressures, institutional imperatives, insecurity and competition, which potentially can subvert NGO ideals (Cooley and Ron 2002).

Such phenomena are equally present in the conservation NGO field, the political economy of which is arguably distinct in at least two ways. On the one hand, NGOs have been prime receivers of post-Rio green financing, becoming convenient alternatives, or rather even outsourced flexible delivery mechanisms, to poor or downsized state arrangements. On the other hand, NGOs remain instrumental in setting up new funding schemes and shaping and channelling contemporary global conservation financing and flows. Both roles, as funding receivers or as gatekeepers, recently have been problematised, not the least being in the public space.

Naomi Klein and others have decried how 'Big Green' relies on capital investments in publicly traded securities in energy, materials and mixed assets. Environmental groups, she argues, have become 'part-owners' of industry, furthermore proposing 'false solutions' and dead ends under the banner of 'constructive engagement'. Specifically, Klein listed conservation organisations as the Nature Conservancy with a USD1.4 billion endowment, trumping the Wildlife Conservation Society (WCS) with only a USD377 million endowment, and WWF US with a mere USD195 million. Klein's message (2013) to Big Green is clear: 'cut your ties with the fossils, or become one yourself'.

Reporter Oliver Steeds wrote about 'Conservation's Dirty Secrets', whereas Canadian journalist Cory Morningstar spoke of the 'non-profit industrial complex' and 'its role to undermine, marginalise and make irrelevant, the People's Agreement'.[3] Alliances were being made with the powerful, not the dispossessed, his narrative goes. Morningstar ended by noting how 'groups who continue to protect such interests must be considered complicit in crimes against humanity'. Mainstream conservation organisations were, noted by another observer, to be considered 'perpetrators' and 'worst offenders' not only limiting their work to trivialities but also in effect 'legitimising plunder' (Barker 2009). Critical voices from the global South equally see NGO power in opposition to people's power. Arundhati Roy questioned the innocence of NGOs speaking of an 'increasingly aggressive system of surveillance of increasingly hardening States' (Roy 2012).

In Bolivia, NGOs have been accused of co-opting and making 'hungry' clients out of indigenous organisations and networks through soft dominance. The NGOs, the country's Vice President argued, were not nongovernmental but organisations of *other* governments (Linera 2012: 30). The language used is remarkably forceful, stating it appears that green civil society was becoming uncivil, nongovernmental was becoming governmental or, even worse, corporate.

From this perspective, the acronym 'NGO' might be recast as 'no good organisations', 'nongoverned organisations', or even 'nature-grabbing organisations' after surveying the recent literature on green grabbing (Corson and MacDonald 2012; Fairhead et al. 2012). Moral resentment (Fassin 2013) that NGOs were siding with power, having 'eaten of the forbidden fruit', threatened the underlying ethical constitution of conservation NGOs. Their very social contract was under fire.

Critical scholarship equally has debunked the 'doing good' narrative, replacing it with less flattering depictions of bigger, bureaucratised and capitalised organisations (Brosius 1999; Escobar 1998). Where NGOs may challenge the state, they may equally offer convenient stop-gap measures to neoliberal policy change (Lewis and Kanji 2009: 18). Conservation NGOs in this vein of thought no longer offer a moral alternative, but they either are subdued by state operations or are caught up in mundane struggles over rights, property, accumulation and redistribution.

A prime example, a decade ago, came from anthropologist Mac Chapin (2004), who created a storm in the conservation community by publishing an article entitled 'A challenge to conservationists'. It was a portrait of three conservation 'giants'—that is, the WWF, Conservation International (CI) and The Nature Conservancy (TNC)—having grown 'extremely wealthy and large', while abandoning earlier commitments to collaborate with indigenous people. The attitude of conservationists was, Chapin (2004: 21) noted, that 'they have the money and they are going to call the shots'. He described the spectacular growth of NGO funding, money dependence and the creation of NGO 'gatekeepers'.

The anthropologist, in a sense, spoke truth to power, a position normally occupied by NGOs themselves. The WWF responded that Chapin's analysis was 'peppered with inaccuracies', yet also resolved to take corrective measures (Roberts and Hails 2005). A member of CI spoke of the article as being 'fraught with errors and unsubstantiated assertions'. Nonetheless, it pointed to friction and tensions arguably omnipresent across the NGO scene.

Such critiques undermined narratives of conservation NGOs as vectors of the good (against power), to portrayals of NGOs as creatures of power and hegemony (Corson 2010). Several NGOs were no longer only local and popular but global and elitist according to Holmes (2011). Such 'grown-up' NGOs have, put in somewhat colloquial terms, moved away from the innocence of green youth to become seasoned business-driven and political, if depoliticised, movers and shakers. Igoe and Brockington (2007: 439), for example, note how big conservation NGOs control 'billions of dollars, employing tens of thousands of people worldwide, and adopting increasingly corporate strategies, organisation and cultures'. The moving of professionals from NGOs to corporate or governmental jobs are, in these narratives, orders of purity and danger and considered conspicuous. As Holmes (2011: 1) notes, conservation elites interact at conferences and 'the roles of NGOs, corporations, and the state are increasingly indistinguishable'.

The NGOs are considered to be part of 'inter-state and/or national power structures', and 'act increasingly like a morph between transnational corporations and government development agencies' (Jepson 2005: 516). They have become part of a 'transnational capitalist class' (MacDonald 2010: 542). Their failure to achieve conservation on the ground is perceived as a result of their 'generalised global approaches' (Rodríguez et al. 2007: 755). In particular, the critique is made in connection with market-based ('neoliberal') conservation approaches (Igoe et al. 2010). Nonetheless, other 'scaling up' and landscape approaches also are considered as a means of domination, further thriving in a neoliberal environment where conservation NGO boards are increasingly populated by corporate representatives (Corson 2010: 581; Holmes 2011: 6–7).

The argument goes even further in terms of capitalist penetration. Rather than producing alternatives, the 'conservation mode of production' produces images, enclosures and commodities ready for capitalist uptake. Conservation is no longer the bulwark against neoliberalism and the penetration of market ideology, but intimately tied to it (Corson 2010; Igoe and Brockington 2007). Brockington and Scholfield (2010: 554) thus argue about how conservation NGOs are 'constitutive of, and central to, the workings and spread of capitalism in sub-Saharan Africa'. According to them, CI essentially has been involved in displaying how conservation has been capitalised and corporatised in terms of expertise networks, linkages with capital institutions and penetration.

MacDonald (2010: 535–36), for example, emphasises how institutional enclosures have resulted in NGOs 'visibly and legibly aligning ... activities, capacities and objectives with the ideological and material interests of the dominant actors'. Corson (2010: 579), furthermore, argues how this relies on a separation of financing for conservation abroad from the driving forces of consumption back home. Not only are market mechanisms rendered unproblematic but also partnerships with businesses are deemed essential to effective conservation. Organisations, from this perspective, are dependent on and ultimately shaped by resource allocation from an external environment driven by neoliberal premises. Because this challenges the founding narratives of 'doing good', the poor reception is not surprising (Igoe et al. 2010).

Still, have such critiques been overdone? Has critique of the 'ugly' conservationist become comfortably radical, potentially misinterpreting and, in some cases, even closing or undermining potential spaces for social change prompted by NGOs? It is thought-provoking that radical critique may appear side by side with government attacks on the NGO sector. The increasingly restrictive regulatory frameworks put in place to curtail NGOs' action in several countries are a real political challenge. How then do we prevent anthropological analysis from becoming stuck in a wholesale post-developmental critique of ugly conservation, not to say misreading NGO tactics? Need we not broaden the scope of critical analysis? Can the answer be found in the pragmatism of what this chapter's author labels the 'Dirty Harry' narrative, often highlighted in the line of defence by conservation organisations?

The 'Dirty Harry' of Conservation

> Because the environmental issues facing humanity are massive and complex, we believe that the most effective way to find solutions is to work with other organisations, including corporations, governments and other NGOs. It is simply not sufficient to throw stones from the sidelines; we must work together to address the needs of a rapidly growing global population that is dependent on a fragile and already overstressed environment. (Seligmann 2011)

The preceding quote by Peter Seligmann, CEO and cofounder of Conservation International, is from a blog post entitled 'Partnerships for the planet: Why we must engage corporations'. Whereas the 'ugly conservationist' narrative portrays organisations caught up in big finance, corporate engagement and neoliberal flirtations, what I call the 'Dirty Harry' narrative, stresses pragmatic conservation operators in a world of money and power. In reference to the famous detective film starring Clint Eastwood, the criminologist Carl Klockars (1980) conceptualised the 'Dirty Harry problem' as the 'moral dilemma' of whether to use 'dirty means' for 'good ends'. Whereas Dirty Harry, a San Francisco cop, tracks down a serial killer sniping random victims, the 'Dirty Harry' of conservation has become the conservationist transgressing the moral codex of the 'good conservationist' to secure effective conservation results.

The credo of the 'Dirty Harry' narrative is that effective conservation requires discreet engagement, adequate resourcing and positive solutions. From a pragmatic conservationist perspective, increasing funds is a necessity in real-world conservation. Where campaigners remain stuck in green utopia, the 'Dirty Harry's of conservation get their hands dirty by managing big budgets and remaining on speaking and operational terms with both government and industry.

The question is no longer about ideological identification with good conservation values, but *how* to get the job done. Gone are the NGOs as the rebellious outsiders revolting against the system, yet without the means, connections and *sagesse* to make any real change happen. 'Yes, there is money involved' and 'yes, there are contacts with corporate and government financing', conservationists may retort. 'Yes, conservation involves professionals and technical debates by professional staff members, rather than only volunteers', they might add. What one on hand, may appear as a huge amount of resources as per observations of a 'conservation industry' is, from this perspective, a question of inadequate conservation finance dwarfed by other budget lines.

David Cleary, an anthropologist working with The Nature Conservancy, characterised Chapin's (2004) early critique as 'incomplete, naive, and overstated caricature of a complex reality'. Instead, he argued, conservation organisations need big budgets to scale-up activities and operate against power behind the scenes, passing intelligence to campaigners, rather than aligning themselves with the poor in public.[4] Should anthropology, in this sense, not get over its structural naiveté and come to terms with the contemporary realities of the 'Dirty Harry's of conservation?

Many anthropologists have, indeed, pursued this pragmatic strategy through intimate engagements within conservation NGOs from local-level conservation board membership to active involvement as staff or advisors. Some have contributed to public debates around conservation policy and practise, while others have remained involved in 'internal' discussions. Ranging from critique at the margins to mainstream policy support, pragmatic engagements in practise reveal high levels of complexity, problematising not only what NGOs do but equally how they engage with the state and corporate actors.

In effect, NGOs may support or substitute governments and the corporate sector because they may be financed, resisted or prohibited by them. They may be curtailed or co-opted as they wrestle from within the system. Working around state control of them is part and parcel of smooth NGO operations in many places. Doane (2014) offers a convincing portrayal of the ups and downs of an NGO-supported community conservation initiative Mexico confronted with decentralised authoritarianism of state agencies working against autonomy-inspired conservation. Language and activities often are tailored and constrained to fit distinct spaces of action, not reflect buy-in altogether. States may accept and allow NGO activity in noncontroversial fields, while restricting action in others (Gunte and Rosen 2012).

At stake, from this perspective, is a different NGO culture, not co-opted by capital but involving pragmatic operators working the system and market-based dynamics to build solutions from within rather than shining brightly from the outside. Nonetheless, such pragmatism comes with a cost. Still, rather than jumping to conclusions about 'nature unbound', of natural relations subjected to market dynamics, what studies in fact reveal are how growing organisations are being unbound and increasingly subject to the marketplace of conservation finance. This, in part, appears as legitimacy framed as 'value for money' rather than only value per se.

As former Director of the WWF, Jim Leape, noted in a 2012 public lecture in Geneva, organisations like the WWF 'help crack the challenges, they [governments] can't crack alone'. He mentioned how his brother, reportedly working for Greenpeace, would complain about how they would 'open the door and WWF would enter'.[5] When campaign organisations stand on the grounds of conservation values, mainstream organisations get their hands dirty to get the job done. Engagement is framed by many environmental organisations as a prerequisite for being effective in today's world (Jepson 2005: 517; MacDonald 2010). Such pragmatism—shifting from protest and activism to professionalism and advice—has long been a trend in mainstream conservation. As a 1976 Editorial in the IUCN Bulletin noted:

> While it may be repugnant for conservationists to divert their precious energies from conservation proper, or to be involved in development at all. … [U]nless we are involved much (if not all) that we have achieved in the past and hope to achieve during the coming years, will be destroyed by the efforts to survive of millions of poor and hungry—helped only by biologically prodigal development on the one hand and socially naive conservation on the other, and therefore not helped at all.[6]

As the initial Morges Manifesto (1961: 2–3) noted, 'success would depend … on winning the respect and backing of other interests, which must not be overlooked or antagonised'. Indeed, from Rio and onward, the sustainability 'bond' cemented NGOs as partners of action in a new pledge: 'No development without sustainability; no sustainability without development' (Sachs 2010: 28). From going against development, conservation NGOs had become efficient vehicles to achieve it.

Nonetheless, whether such nonantagonism ultimately is successful or in practise undermines NGO independence and conservation impact is under constant debate. Interestingly, as global trends of NGO engagement with the corporate sector have increased, CEOs reportedly perceive a declining significance of NGOs as trendsetters.[7] Whereas 27% of CEOs in 2007 listed NGOs among their top three in terms of stakeholder groups influencing approaches to sustainability, this had fallen to 15% in 2013. Is the agenda-setting edge ability of NGOs eroding while they have grown in terms of size, budgets and power? Answering this question is beyond the scope this chapter. Yet, clearly part of the answer will rely on careful assessments of the shifting politics and practises of conservation NGOs (Robinson 2012). The following section explores such questions through a case study of conservation NGO activity in the Peruvian Amazon.

Doing Good in the Peruvian Amazon

'We are like Don Quixote', a conservationist exclaimed following yet another meeting about setting up a biosphere reserve in the central jungle area of Peru in early 2008. In the context of massive development pressures, he joked about the group of conservation players around the table on a noble quest to change the world. The image had a ring to it. The tireless group attempted to promote a biosphere reserve project, question the expansion of the electricity grid into a municipal nature reserve and otherwise get green issues on the agenda. Whereas Don Quixote regains sanity, these conservationists arguably pursued high goals and ideals against challenging development odds. They were, in the narrative sense, agents of the 'good'. Many were NGO employees, others former ones. They were, I would argue, not merely acting through personal conviction, but also reenacting NGO identity narratives of environmental avant-gardism and commitment to the public good.

A quarter of a century prior to his critique of conservation NGOs, Mac Chapin, then a Latin American advisor at United States Agency for International Development (USAID), facilitated one of the agency's first social and environmental impact assessments of a development project. Civil society critique, from anthropologists to indigenous organisations and their supporters had thrown into question government plans for road expansion and frontier settlements in the Peruvian Amazon (Smith 1982). The USAID financing and conditionalities eventually led to protected area creation, land titling and sustainable forest management projects. This reoriented project space, with only nominal state support, carved out a distinct managerial vacuum ready for NGO support. The *Fundación Peruana para la Conservación de la Naturaleza* (FPCN)—that is, the Peruvian Foundation for the Conservation of Nature, later recognised as *ProNaturaleza* (created in 1984)[8]—was the first national conservation NGO founded by key Peruvian conservationists[9] in response to the limitations of state action (Husock 1997). Protected area financing was not a public funding priority, and conservationists set up the foundation to receive international donor support. Whereas foreign support for conservation in the late 1970s was 'never more than 8% of state contributions' (Dourojeanni and Ponce 1978: 19), these figures would be reversed in the coming decades.

Shawn Miller (2007: 194) spoke of 'Latin American nature' generating 'an unusually large share of first-world environmental anxiety'. Such anxieties after Rio translated into a managerial reconfiguration of problems as 'technical', centred around renewable resources (Chatterjee and Finger 1994; Hajer 1995), consolidating the role of NGOs as mediators between the resource-rich North and the biodiversity-rich South. The NGOs were among the key supporters, fundraisers and beneficiaries of the managerial paradigm and new finance mechanisms.

Whereas debt swaps and international finance during the late nineteenth century had fuelled the agricultural frontier in the Peruvian jungle, a century later Northern finance would (attempt to) render the region green.[10] The FPCN 'became the means for international donors ... to give meaning to Peru's protected areas' (Husock 1997: 4). Tropical forests predominated hotspot lists, and Oxapampa fit the bill, forming part of the tropical Andes—'the richest and most biodiverse hotspot on earth ... the global epicentre of biodiversity' (Mittermeier et al. 1999: 69).

The NGO actions, during this period, were about filling the gap of an absent state. The FPCN's aim was to become a 'centre of excellence' in conservation programmes and biodiversity projects (Husock 1997: 3). Founding logics were, in the language of this analysis, based on 'doing good' in technical terms by initially replacing the absent state by providing protected area management. It was not an activist NGO, but a foundation structure managed by conservation professionals and former government staff, essentially developing a kind of parallel public structure. Peruvian industrialists were soon to be members of the board (Husock 1997: 5). Not only did the NGO itself manage parks, but it also organised training and workshops on management planning. Although no formal mandate was given, FPCN initially had a 'gentleman's agreement' and later, through donor pressure, obtained a 'compact of cooperation' (1986–1989), becoming like a de facto state-protected area authority (Ibid: 4–5).

One of its first major operations, with USAID and TNC support, was the direct management of the Yanachaga National Park in 1987. By 1990, the NGO was administering all the protected areas in the province. A few years later, the NGO supported three-quarters of the national protected area system, specifically in charge of 15 of them.[11] The FPCN even cosponsored the incorporation of a chapter on natural resources in the 1993 Constitution.[12] Obviously, this was much more than 'doing good'.

It was NGO governance par excellence, replacing the state yet without having to deal with cumbersome national politics. Like many other NGOs, expansion relied heavily on a project economy, eventually becoming part of the multi-country 'Parks in Peril'[13] TNC flagship programme 'transforming "paper parks" into functional protected areas' (González and Martin 2007: Foreword). The TNC had, by then, grown from a small group of ecologists in 1950 to a major conservation player. By 2005, TNC had more than USD3.7 billion in assets and an annual revenue of USD800 million (Birchard 2005). For the Peruvian NGO, this entailed the emergence of a specific process of priority-setting that reflected managerial prerogatives of the organisation. In the Selva Central, this went from park-oriented support in the period from 1991 to 1997 to a second phase supporting two other protected areas in the province.

Although much of the preceding, in isolation, illustrates the emergence of a conservation NGO doing good fuelled by international support, there was more to the story. Global conservation finance fuelled NGO-driven problem analysis rendering conservation systems the locus of environmental problem solving. In the following years, a series of projects around biological diversity, forests and fauna nurtured a distinct green vision of Oxapampa and its management needs. This led to an emphasis on natural forest management and protected areas, with less emphasis on agricultural production, soil conservation and contamination issues.

The Ugly Conservationist?

The fuelling of green managerial power had social effects. The prior tying together conservation and social rights, which had mobilised protests and enabled protected area creation in the first place, was being undone. Indigenous communities, which had fought against logging and road expansion, were distanced from protected area management. Where indigenous organisations and their supporters had supported indigenous communal reserves, conservationists put a greater emphasis on conventional state-driven management responses (Larsen 2015). The NGO staff channelled the bulk of the funding to the uninhabited national park. 'We had to prioritise', as one conservationist explained (pers. comm.). As a result, NGO priorities shifted in 1997 when the government reclaimed direct management of the park. The state agency laid off all forest guards hired by the NGO, triggering a shift in role and perspective.

As the state focussed all its energy on the park, it left a residual project space for the NGO, by then renamed ProNaturaleza. Project attention was shifted to other protected areas from which the state still remained 'absent'. The NGO initiated capacity-building support and studies in the San Matías-San Carlos Protection Forest and the Yanesha Communal Reserve. As Steven Sampson (2002) noted, 'project life is a world with a premium on abstract knowledge, by which power accrues to those best able to manipulate key symbols and concepts'. Just like capital ventures are expected to generate yearly profits, conservation initiatives are expected to generate added value and success stories (Sachedina et al. 2010: 25). New capacity-building roles, and, most important, new projects were defined to target community organisations and local authorities. Nonetheless, this project economy, like many others, also had unintended consequences.

By 2006, ProNaturaleza, promoting the idea of a 'Central Selva Biosphere Reserve', had lobbied to place the proposal on the agenda of a decentralised Ministerial Council meeting. Even though it received support from Alejandro Toledo, the country's president, alternative proposals stressing the need for more emphasis on the indigenous perspective emerged, eventually leading indigenous organisations to withdraw support.[14] 'ProNaturaleza received a lot of money, dollars, in our name', one informant mentioned. 'Money was not invested here, where it should have been, it went to another site. … [T]hey would continue as usual, with millions of dollars … [but] our situation wasn't changing', another added (pers. comm.). Mistrust went beyond an individual project and was echoed by both indigenous leaders and local authorities on multiple occasions.

Although NGO work was appreciated and solicited, NGOs doing conservation good were seen equally as reaping benefits from environmental problems and leaving little behind. Critique was not the least rehearsed and was mobilized as part of local politics. The NGOs were easy targets and scapegoats for action, but equally visible manifestations of deep-running inequities. 'The NGOs see us as children', one indigenous leader stated. 'You [indigenous] aren't able to talk directly to the state … and the state equally thinks that as indigenous peoples, we aren't yet civilized. … The NGOs and the church should be in charge, so that the *indios* get civilised', he continued. One indigenous *comunero* (i.e., commoner) even criticised his own leaders using this image of NGOs: '[T]he presidents live big style, they have converted themselves into another ProNaturaleza'. Some NGOs representations featured four-wheel drive project vehicles, high salaries and office space. They were, in one sense, 'ugly' victims of their project success.

In the meantime, international support for conservation in the region was dwindling. Major projects came to an end, people were laid off and office facilities left at their bare minimum. The organisation had gone through a major financial crisis. Funding was gone, except for some minor projects, and the NGO was no longer offering big-scale solutions to forest and protected area management. Big conservation, it seemed, was only a shadow of its past. The NGO advisors, instrumental in supporting protected areas, had become consultants for local authorities, moved elsewhere or joined other institutions. Furthermore, despite long-standing criticism, indigenous organisations were busy setting up their own NGOs. By 2013, the indigenous federation of the Yanesha, the *Federacion Comunidades Nativas Yanesha* (FECONAYA), had set up its own organisation. Whereas indigenous political organisations relied on voluntary involvement, NGOs required 'professionals', mirroring the evolution of the national NGO culture during the 1980s and 1990s.

Another indigenous organisation, Asociación para el Manejo y Conservación de la Reserva Comunal Yanesha (AMARCY), with a protected area comanagement mandate for the Yanesha Communal Reserve, was equally contemplating formal recognition as an NGO. As Yanesha leaders went from critique of NGOs to setting up their own, it was all about activating the power of a social form they had experienced. It also involved adopting practises and language of the *ingenieros* (i.e., engineers), characterised by ambitious project proposals, reporting and PowerPoint presentations. As one Yanesha expressed it: '[I]f an NGO can fetch one million soles without consultation. ... It's time to turn the page and undertake direct negotiations'. Another Yanesha told this author, 'We want our own NGO'.

One of their first sources of funding had been project development resources from an oil company. Yet leaders also were coming to terms with the 'ugly' side of project funding and power politics. An indigenous leadership crisis erupted in 2013 around the management of project resources. Funding from the oil company remained a major source of contention. Management of the NGO was at the heart of the matter and dynamics were strikingly similar to the narrative of 'ugly conservation'. Still, the case also illustrates the limitations of stereotypical dichotomies between big conservation and local communities. Whereas 'ugly conservationist' elements appeared at different moments, use was transitory, rhetorical and part of complex local politics. Rather than a stable property of conservation NGOs, it even reappeared in the context of indigenous NGO creation. At stake were complex interplays between state politics, indigenous representative organisations, corporate players and shifting funding schemes.

Is 'Dirty Harry' in the Amazon?

> Through our role as a technical advisor, we aim for efficient socio-environmental management allowing for spaces of communication between the local population and the executing companies of extractive and construction activity. – *ProNaturaleza* (Dourojeanni et al. 2012: 195, translated from Spanish)

During return visits to Central Selvl in 2010 and 2013, earlier project activities by ProNaturaleza had been sharply reduced. Nevertheless, programme activities in other parts of the country had grown considerably.

The NGO was increasingly active in supporting environmental monitoring programmes for the oil and gas bonanza in the Peruvian Amazon (Finer et al. 2008). In 2001, ProNaturaleza had already begun to organize a community-based monitoring programme in the Camisea gas fields. The schemes were considered successful in terms of early detection of problems and preventing local conflicts in a 'silent and routine form' (Dourojeanni et al. 2012: 152). In a decade, such schemes had grown to become a major programme activity.

By 2013, the NGO was under contract to work on community monitoring with seven companies (five oil and two mining). The organisation described itself as a 'technical advisory body' defining one of its four strategic lines of activity as 'searching for conservation and good land use in places where energy and road infrastructure projects are implemented through the reconciliation of corporate and local community interests' (ProNaturaleza 2013: 9). Monitoring was the technical means to achieve this. Such win–win language around the effectiveness of technicality illustrated a distinct NGO positionality 'outside' the political battlefield. The director of the NGO would present the objective of involvement in monitoring as securing high standards of environmental quality in extractive projects, reducing levels of social conflict, maintaining fluid communication along with knowledge transfer and empowerment of communities (Dourojeanni et al. 2012: 7).

Participatory and community-based monitoring schemes were evaluated as 'overwhelmingly beneficial' in terms of offering 'additional guarantees', empowerment and an increased understanding (Ibid: 147). Companies were even considered to benefit 'more' from the scheme in terms of 'avoiding conflicts, and to their own surprise, help them in avoiding otherwise grave accidents' (Ibid: 151). Whereas the NGO in its early years aimed at transforming paper parks into effective management, the role in extractive industry monitoring increasingly involved becoming a trusted, efficient and competent service provider (Ibid: 186).

Had conservationist practise gone from civil society influencing state practise to become part of corporate project governance? Was it 'Dirty Harry' conservationism in practise? The monitoring scheme, practitioners argued, specifically involved empowering indigenous communities to set up their own monitoring bodies. Protests and denunciations were, from the NGO perspective, seen as 'exaggerations' and distortions provoked by special or ideological interests such as the nationalist left (Dourojeanni et al. 2012: 16). 'Doing good' was not about open campaigning and advocacy against oil in certain areas and alternative land use, but about demonstrating that the job could be done 'properly'. It was about getting hands dirty to secure clean technologies.

In Oxapampa, where ProNaturaleza had been a major conservation pioneer, oil exploration was taking place within or next to the protected areas the organisation had established. Local authorities were confronted with major challenges to respond effectively to potential social and environmental impacts (Larsen 2011; Larsen and Gaspar 2012). Yet, despite having a national programme on oil monitoring, the NGO remained surprisingly silent on the topic of exploration activities compared to other NGOs and protected area authorities. Lack of project funding appeared to be the major reason, illustrating the fragile nature of NGO positioning on major development challenges. The NGO staff were well aware of the conflicts at stake; however, they also operated in project economy finance monitoring interventions elsewhere. Independent opinion was not absent per se, but it was channelled through and, in some respects, replaced by positionality determined by the political economy of project funding and was framed as technically sound project advice. Because NGOs are made up of various undertakings, projects in some respects made up the public face of NGOs de facto determining positionality.

While this chapter was being reviewed, the NGO (described as an 'environmental consultancy' by one observer) was hired by the oil company, Pluspetrol, to work in the neighbouring Concession 108 to set up a monitoring programme and citizen environmental vigilance programme. The USD200,000 project reported on the NGO website aimed to initiate a programme with 'adequate local representation' contributing to 'improving the compliance of Pluspetrol in relation to commitments made in the environmental management plan'.[15] In a controversial and politicised exploration operation triggering significant protests, indigenous leaders and NGO staff, who previously challenged corporate oil, now offered local community members technical training to monitor operations and 'improve compliance'.

The point here is that careful analytical attention is needed to consequences and the social effects of 'Dirty Harry' approaches to explore the conditions and implications of engagement. This challenges analysts to look beneath the narratives of strategy documents, and to work with a more fluid notion of NGO action crafted around actual practises of engagement and real-life project economies.

Discussion: Approaching NGO Complexity

ProNaturaleza and other NGOs have at different moments been viewed as doing good, representing the ugly and engaged in 'Dirty Harry' practises. Creation of NGOs was, for example, largely framed around 'doing good' by substituting or supporting an absent state. Fuelled by international finance, the NGOs grew rapidly, yet not without reactions. Local criticisms of NGO projects and distance to communities echoed the master narrative of 'ugly conservation'. Finally, NGO repositioning as a technical advisory body to extractive industry projects evoked the master narrative of 'Dirty Harry' conservation. Still, such master narratives provide only a very partial picture of the complex dynamics involved.

First, the historical perspective planned in the Peruvian case study revealed major organisational changes within a short timespan. Conservation NGOs are rapidly evolving rather than stable organisations and fields of activity. Even though it is well recognised that NGOs are diverse (Igoe et al. 2010: 6), far more attention needs to be paid to (1) the internal heterogeneity, (2) the highly unstable terrains of conservation NGOs, and (3) the evolving conditions over time.

Second, the case points to the significance of changing waves of donor finance. The NGOs' roles were closely tied to evolving conditions from levels of government involvement, resource politics and specific finance opportunities. The NGOs may indeed be conceived not as self-contained, but as entities that rely on external environments: 'It is this dependence on an external environment that not only makes the control of organisational behaviour possible but almost inevitable as NGOs need to be appropriately responsive to that environment to assure continued access to the resources they need to survive' (MacDonald 2010: 534).

Nongovernmental organisations are, in this sense, not organisations with a predefined agendas, but rather made up of embedded projects with implicit normative positioning. This raises questions not just about what NGOs do but also fundamentally about what they are. As Pinzas et al. (1997: 6) has noted: 'Many of the NGOs are actually not organisations, but rather a loose collection of projects run by a single agent which are not interlinked or mutually supportive—if one fails, the remainder continue; the unsuccessful die and the successful grow'.

Although much attention is directed at NGOs themselves, this suggests the need for far more analytical attention to the surrounding social and political processes constraining or enabling specific forms of action. This is arguably at the heart of the literature situating NGO practise in the wider context of a political economy framed around neoliberalism as well as literature emphasising the determining role of donor priorities. Despite that, this also needs to go beyond the totalising gaze of neoliberal critique and to address the broader picture of contextual dynamics and politics.

Third, the case reveals the relevance of tracking NGO positionalities and relationships, both over time and across space, so as to capture their situational and contingent nature. Narratives stressing powerful NGOs tied into the corporate-status nexus reveal only one side of the coin. Brockington and Scholfield (2010), for example, note how conservation NGOs only support 15% of the protected area network in sub-Saharan Africa, influence being unevenly distributed among specific regions, countries and definite sites. Changing relations of cooperation, partnership and competition are central features of NGO activity (for the Guatemalan example, see Grandia 2010). Most NGOs are in a constant process of transforming associations and repositioning themselves in relation to internal or external factors such as international funding priorities (Grandia 2009). Rather than merely denouncing (lack of) interaction as 'ugly conservation' or displaying 'Dirty Harry' strategies, the question is whether and how strategies of 'rapprochement' are transforming conservation priorities and dynamics.

Critical questions are under what conditions and with what results, conservation NGOs engage with the corporate sector, government and international finance. This presents NGO studies with the challenge of capturing how shifts in financing and projects enable or foreclose specific NGO positions and their ability to actually influence change. The rise and fall of NGO power, staffing and field presence over time, because of changing flows of finance, are fundamental dynamics in this respect. In a recent review, Robinson (2012: 975) found little empirical evidence of NGO engagements with the corporate sector leading to better conservation results, even concluding not to 'rely on corporations to meet conservation goals'. Portraying the variety of relations from dialogue to philanthropy and collaboration, his analysis specifically challenges win–win optimism based on voluntary measures and Corporate Social Responsibility (CSR).

Concluding Remarks: Rethinking the Anthropology of Conservation NGOs

Academic attention to the nature and the role of conservation NGOs is growing in parallel with the growth and transformation of the organisations themselves. Conservation anthropology has long interrogated the 'local' slot of conservation practise through problem analysis, field activities and policy work. The anthropology of conservation NGOs in turn has proved to be a productive reality check that stresses the power and politics of conservation practise. Critical analysis, for example, points to the growing importance of corporate partnerships and state imbrications of NGOs forming part of sustainability problems and power fields, rather than merely observing challenges from the outside. The discrepancies between NGO self-representation, and the tactics of professionalised bureaucracies of the big and capital-intensive conservation machineries, is a constant source of societal interrogation; need anthropological critique stop there? Has such critique of power become comfortably radical and, at times irrelevant, for the complex set of social battlefields making up nongovernmental conservation action?

Critical analysis may be more or less welcomed by the organisations themselves (Benjaminsen and Svarstad 2010; Igoe et al. 2010); however, it is arguably essential to foster a constructive debate about the role and contribution of civil society in addressing the daunting environmental challenges of our times. How is a critique of conservation NGOs maintained without throwing the baby, that of nongovernmental or civil society action, out with the bathwater? How is anthropology neither comfortably radical at a distance, nor comfortably operational from within the 'social slot'? How are critical tools and findings better mobilized within conservation organisations to challenge working assumptions and modalities?

This chapter argues that master narratives around the good, the ugly or the 'Dirty Harry's of conservation continue to frame the contours of the debate, yet provide only partial insights into the complex realities of conservation NGOs. Between 'the more NGO activity the better' and a wholesale critique, a critical middle ground of anthropological analysis is emerging. This middle ground is fundamental both to capture problematic spaces and to alter institutional forms and practise worthy of anthropological exploration (Lockyer and Veteto 2013). This may avoid the traps of meta-narratives equalling BINGOs with big, bad and business-minded or, conversely, idealising small and beautiful Southern conservation efforts.

The problematisation of NGOs has opened new avenues of investigation into the changing conditions, uneven nature and evolving practises at stake. Conservation NGOs, like many other NGOs (Lewis and Kanji 2009), face choices about who to engage with, where to prioritise activities and how to engage with the wide range of stakeholders impacting and benefitting from conservation. In practise, a panoply of 'hybrid' conservation NGOs, defying conventional dichotomies, are found with varying degrees of international financing, local ownership and corporate involvement. This also concerns everyday decisions about how capacity-building is undertaken, how collaboration is built, and the extent to which vibrant local institutions are bolstered (Rodríguez et al. 2007: 756). The question is no longer whether NGOs (can) make a difference, but indeed 'what' difference they make as well as when, where and how they can make it.

What in one context might appear as professionalisation, bureaucratisation and monolithic power is countered by changing transformative politics, local alliances and internal heterogeneity in other cases. This entails a less essentialist and more dynamic notion of NGOs rather than attributing specific properties to them. Critical analysis need not limit itself to denouncing global power, corporate board membership and partnership language, but it might further explore the detailed trajectories of conservation impact, policy influence and long-term effects of such engagements. Considering that data on the effectiveness conservation and development NGOs often is scarce (Lewis and Kanji 2009; Robinson 2012), anthropological descriptions that pay attention to evolving NGO trajectories and shifting terrains of intervention, and their social effects, are among the most critical building blocks not only to the field of NGO studies but equally so for NGO practise at large.

Notes

1. This chapter was first presented as a paper at a panel entitled "The Anthropology of Conservation NGOs", held in Chicago in December 2013. Thanks to other participants for stimulating discussions as well anonymous peer reviewers for insightful comments and suggestions.
2. Group mail received on 5 June 2013.
3. http://wrongkindofgreen.org/about-us/
4. http://www.thefreelibrary.com/More+responses+to+%22A+Challenge+to+Conservationists%22.-a0130057621. Accessed on 14 November 2013.
5. 'Sustainable Development: The Agenda After Rio+20', Conference given at IHEID, October 9, 2012.
6. Editorial Comment IUCN Bulletin, New Series 7(1), January 1976.
7. http:/www.theguardian.com/global-development-professionals-network/2013/sep/20/ngos-no-longer-set-agenda-development. Accessed on 10 May 2014.
8. In Peru, Pro Defensa de la Naturaleza (PRODENA) was the first NGO to deal with environmental matters particularly regarding natural resource management. Several successive splits from this organisation gave birth

to *Fundacion Peruana para la Conservacion de la Naturaleza* (known as ProNaturaleza), *Asociacion Peruana para la Conservacion de la Naturaleza* (APECO) *Asociacion de Ecologia y Conservacion* (ECO) and other organisations that triggered the development of the environmental movement (Soria unpublished).

9. The three founders, Marc Dourojeanni, Carlos Ponce and José Rios were all La Molina Agrarian University faculty in the 1970s. Dourojeanni and Ponce had held key positions in the Directorate of Flora and Fauna.
10. Where the indebted Peru in the nineteenth century had offered British bondholders land in the Oriente, twentieth-century debt swaps offered to the German government and other protected area designation and management plans.
11. http://www.pronaturaleza.org/pronaturaleza-cumple-25-anos/. Accessed on 10 January 2011.
12. By 1995, FPCN had become ProNaturaleza, who cosponsored the constitutional element with an environmental lawyer NGO (*Sociedad Peruana de Derecho Ambiental*, SPDA).
13. This, mainly USAID, funded the TNC programme covering Latin America and the Caribbean ran for 17 years with projects for some USD104 million.
14. The biosphere reserve proposal, later renamed the Oxapampa Ashaninka Yanesha Biosphere Reserve, would eventually take off again with greater involvement and protagonism of the indigenous federations. Largely supported by NGO actors in the province, it would be approved in 2010.
15. http://pronaturaleza.org/wp-content/uploads/2013/nuestro-trabajo/promocion-de-la-responsabilidad-socio-ambiental/Pasco-Junin.pdf. Accessed on 27 April 2015.

References

Bamberg, M. 2005. Master Narratives. In *Routledge Encyclopedia of Narrative Theory*, ed. D. Herman, M. Jahn, and R. Marie-Laure, 287–288. Abingdon: Routledge.

Barker, Michael. 2009. When Environmentalists Legitimize Plunder. *Swans Commentary*. http://www.swans.com/library/art15/barker12.html. Accessed 18 Sep 2013.

———. 2010. An Interview with Daniel Faber: Foundations and the Environmental Movement. *Counterpunch*. http://www.counterpunch.org/2010/09/13/foundations-and-the-environmental-movement/. Accessed 13 Sep 2010.

Benjaminsen, Tor A., and Hanne Svarstad. 2010. Questioning Conservation Practice—And Its Response: The Establishment of Namaqua National Park. *Current Conservation* 3 (3): 8–9.

Birchard, Bill. 2005. *Nature's Keepers: The Remarkable History of How the Nature Conservancy Became the Largest Environmental Organization in the World*. 1st ed. San Francisco: Jossey-Bass.

Brockington, Dan, and Rosaleen Duffy. 2010. Capitalism and Conservation: The Production and Reproduction of Biodiversity Conservation. *Antipode* 42 (3): 469–484.

Brockington, Dan, and Katherine Scholfield. 2010. The Conservationist Mode of Production and Conservation NGOs in Sub-Saharan Africa. *Antipode* 42 (3): 551–555.

Brosius, Peter. 1999. Green Dots, Pink Hearts: Displacing Politics from the Malaysian Rainforest. *American Anthropologist* 101 (1): 36–57.

Brosius, Peter, and Lisa Campbell. 2010. Collaborative Event Ethnography: Conservation and Development Trade-Offs at the Fourth World Conservation Congress. *Conservation and Society* 8 (4): 245–255.

Chapin, Mac. 2004. A Challenge to Conservationists. *World Watch Magazine* 17 (November/December): 17–31.

Chatterjee, Pratap, and Matthias Finger. 1994. *The Earth Brokers: Power, Politics and World Development*. London/New York: Routledge.

Cooley, Alexander, and James Ron. 2002. The NGO Scramble: Organizational Insecurity and the Political Economy of Transnational Action. *International Security* 27 (1): 5–39.

Corson, Catherine. 2010. Shifting Environmental Governance in a Neoliberal World: USAID for Conservation. *Antipode* 42 (3): 567–602.

Corson, Catherine, and Kenneth lain MacDonald. 2012. Enclosing the Global Commons: The Convention on Biological Diversity and Green Grabbing. *The Journal of Peasant Studies* 39: 263–283.

Doane, Molly. 2014. From Community Conservation to the Lone (Forest) Ranger: Accumulation by Conservation in a Mexican Forest. *Conservation and Society* 12 (3): 233–244.

Dourojeanni, Marc Jean, and Carlos F. Ponce. 1978. *Los parques nacionales del Peru*. Madrid: Instituto de la Caza Fotográfica y Ciencias de la Naturaleza.

Dourojeanni, Marc, Luis Ramirez, and Oscar Rada. 2012. *Indígenas, campesinos y grandes empresas: Experiencias de los Programas de Monitoreo Socio-Ambiental Comunitarios*. Lima: ProNaturaleza.

Escobar, Arturo. 1998. Whose Knowledge, Whose Nature? Biodiversity, Conservation, and the Political Ecology of Social Movements. *Journal of Political Ecology* 5: 53–82.

Fairhead, James, Melissa Leach, and Ian Scoones. 2012. Green Grabbing: A New Appropriation of Nature? *The Journal of Peasant Studies* 39 (2): 237–261.

Fassin, Didier. 2013. On Resentment and Ressentiment: The Politics and Ethics of Moral Emotions. *Current Anthropology* 54 (3): 249–267.

Finer, M., Clinton N. Jenkins, Stuart L. Pimm, Brian Keane, and Carl Ross. 2008. Oil and Gas Projects in the Western Amazon: Threats to Wilderness, Biodiversity, and Indigenous Peoples. *PloS One* 8: e2932.

González, Ana Maria, and Angela Sue Martin. 2007. *Partners in Protected Area Conservation: Experiences of the Parks in Peril Programme in Latin America and the Caribbean*. Arlington, VA: The Nature Conservancy.

Grandia, Liza. 2009. Special Issue Between Bolivar and Bureaucracy: The Mesoamerican Biological Corridor. *Conservation and Society* 5 (4): 478–503.

———. 2010. Silent Spring in the Land of Eternal Spring: The Germination of a Conservation Conflict. *Current Conservation* 3 (3): 10–13.

Gunte, Michael M., and Ariane C. Rosen. 2012. Two-level Games of International Environmental NGOs in China. *William and Mary Policy Review* 3: 270–294.

Hajer, Maarten. 1995. *The Politics of Environmental Discourse: Ecological Modernization and the Policy Process*. Oxford: Clarendon Press.

Holdgate, Martin. 1999. *The Green Web: A Union for the World Conservation*. London: IUCN & Earthscan.

Holmes, George. 2011. Conservation's Friends in High Places: Neoliberalism, Networks, and the Transnational Conservation Elite. *Global Environmental Politics* 11 (4): 1–21.

Husock, Howard. 1997. *Dealing with a Changing Environment: Fundacion Peruana para la Conservacion de la Naturaleza*. Cambridge, MA: Kennedy School of Government, Case Programme.

Igoe, James, and Dan Brockington. 2007. Neoliberal Conservation: A Brief Introduction. *Conservation and Society* 5 (4): 432–449.

Igoe, James, Sian Sullivan, and Dan Brockington. 2010. Problematising Neoliberal Biodiversity Conservation: Displaced and Disobedient Knowledge. *Current Conservation* 3 (3): 4–7.

Jepson, Paul. 2005. Governance and Accountability of Environmental NGOs. *Environmental Science & Policy* 8: 515–524.

Khare, Arvind, and David B. Bray. 2004. *Study of Critical New Forest Conservation Issues in the Global South New York*. New York: Ford Foundation.

Klein, Naomi. 2013. The Giants of the Green World That Profit from the Planet's Destruction. *The Guardian* (first published in the Nation). May 3.

Klockars, Carl B. 1980. The Dirty Harry Problem. *The Annals of the American Academy of Political and Social Science* 452: 33–47.

Laidler-Kylander, Nathalie, John A. Quelch, and Bernard L. Simonin. 2007. Building and Valuing Global Brands in the Nonprofit Sector. *Non-profit Management and Leadership* 17 (3): 253–277.

Larsen, Peter Bille. 2011. Municipal Environmental Governance in the Peruvian Amazon: A Case Study in Local Matters of (in)Significance. *Management of Environmental Quality* 22 (3): 374–385.

———. 2013. The Politics of Technicality: Guidance Culture in Environmental Governance and the International Sphere. In *The Gloss of Harmony: The Politics of Policy Making in Multilateral Organisations*, ed. Birgit Müller. London: Pluto Press.

———. 2015. *Post-frontier Resource Governance: Indigenous Rights, Extraction and Conservation in the Peruvian Amazon*. London: Palgrave Macmillan.

Larsen, Peter Bille, and Arlen Gaspar. 2012. *Biosphere Reserves and Extractive Industries: Towards a New Sustainability Agenda, Evidence for Policy Series*. La Paz: NCCR JACS SAM.

Levine, Arielle. 2002. Convergence or Convenience? International Conservation NGOs and Development Assistance in Tanzania. *World Development* 30 (6): 1043–1055.

Lewis, David, and Nazneen Kanji. 2009. *Non-governmental Organizations and Development*. Abingdon: Routledge.

Linera, Alvaro García. 2012. *GeoPolítica de la Amazónia: Poder Hacendal-Patrimonial y Acumulación Capitalista*. La Paz: Vicepresidencia de la República.

Lockyer, Joshua, and James Veteto. 2013. *Anthropology Engaging Ecotopia. Bioregionalism, Permaculture, and Ecovillages*. New York: Berghan.

MacDonald, Kenneth. 2010. The Devil Is in the (Bio)Diversity: Private Sector 'Engagement' and the Restructuring of Biodiversity Conservation. *Antipode* 42 (3): 513–550.

Morges Manifesto. 1961. *We Must Save the World's Wildlife: An International Declaration*. Morges: WWF.

Miller, Shawn. 2007. *An Environmental History of Latin America*, New Approaches to the Americas. Cambridge: Cambridge University Press.

Mittermeier, Russell A., N. Myers, and C.G. Mittermeier. 1999. *Hotspots: Earth's Biologically Richest and Most Endangered Terrestrial Ecoregions*. San Pedro Garza Garcia: Cemex & Conservation International.

Pinzas, T., J. Danziger, and B. Pratt. 1997. *Partners or Contractors? The Relationship Between Official Agencies and NGOs–Peru*. Oxford: Intrac.

ProNaturaleza. 2013. *Memoria Institucional 2012*. Lima: ProNaturaleza–Fundación Peruana para la Conservación de la Naturaleza.

Roberts, Carter, and Chris Hails. 2005. From the World Wildlife Fund (WWF). *World Watch Magazine* 17: 16–17.

Robinson, John. 2012. Common and Conflicting Interests in the Engagements Between Conservation Organizations and Corporations. *Conservation Biology* 26: 967–977.

Rodríguez, J.P., A.B. Taber, P. Daszak, et al. 2007. Globalization of Conservation: A View from the South. *Science* 317: 755–756.

Roy, Arundhati. 2012. Capitalism: A Ghost Story. *Outlook India*. http://www.outlookindia.com/article/capitalism-a-ghost-story/280234. Accessed 26 Mar 2012.

Sachedina, H., J. Igoe, and D. Brockington. 2010. The Spectacular Growth of a Conservation NGO and the Paradoxes of Neoliberal Conservation. *Current Conservation* 3 (3): 24–27.

Sachs, W. 2010. *The Development Dictionary*. London/New York: Zed Books.
Sampson, Steven. 2002. Weak States, Uncivil Societies and Thousands of NGOs: Western Democracy Export as Benevolent Colonialism in the Balkans. In *Cultural Boundaries of the Balkans*, ed. Sanimir Recic. Lund: Lund University Press.
Seligmann, Peter. 2011. Partnerships for the Planet: Why We Must Engage Corporations. *Huffington Post Blog*. Posted: 05/19/2011 12:56 pm EDT Updated: 07/19/2011 5:12 am EDT. http://www.huffingtonpost.com/peter-seligmann/conservation-international-lockheed-martin_b_863876.html. Accessed 24 Dec 2015.
Smith, Richard C. 1982. Dialectics of Domination in Perú: Native Communities and the Myth of the Vast Amazonian Emptiness: An analysis of Development Planning in the Pichis Palcazu Special Project. In *Cultural Survival Occasional Paper 8*. Cambridge, MA: Cultural Survival Institute.
Waldron, Anthony, et al. 2013. Targeting Global Conservation Funding to Limit Immediate Biodiversity Declines. In *Proceedings of the National Academy of Sciences*. Early Edition: 12144–12148.
West, Paige. 2006. *Conservation Is Our Government Now: The Politics of Ecology in Papua New Guinea*. 1st ed. Durham: Duke University Press.

CHAPTER 3

Anthropology of Conservation NGOs: Learning from a Sectoral Approach to the Study of NGOs

Dan Brockington, Katherine Scholfield, and Richard Ladle

Introduction

This chapter draws on a survey of the conservation non-governmental organisation (NGO) sector in sub-Saharan Africa that Brockington and Scholfield carried out in the late 2000s. That project began with a simple curiosity to learn more about the role and impact of smaller conservation NGOs on the continent. We discovered we knew very little about what that sector was like at all. Nor did there seem to be much general knowledge as to its stucture or condition. So, the initial enquiry regarding the

D. Brockington (✉)
Institute for International Development, University of Sheffield, Sheffield, South Yorkshire, UK

K. Scholfield
Synchronicity Earth, London, UK

R. Ladle
Institute of Biological Sciences and Health, Universidade Federal de Alagoas, Maceió, Brazil

P.B. Larsen, D. Brockington (eds.), *The Anthropology of Conservation NGOs*, Palgrave Studies in Anthropology of Sustainability, DOI 10.1007/978-3-319-60579-1_3

work of smaller organisations turned into a larger goal of mapping the entire conservation NGO sector working in sub-Saharan Africa (nearly 300 organisations).

This chapter does not go into the methods used in that survey and their inadequacies; these have been detailed elsewhere (Scholfield and Brockington 2008). Indeed, various aspects of the whole survey are available in several publications (Brockington and Scholfield 2010a, b, c). The purpose here is not to report the results (although we will outline them), but to reflect on what can be learned from treating NGOs as a sector, as opposed to studying individual organisations. By this we mean treating conservation NGOs as being part of a larger group of organisations that share sufficiently common interests, goals, passions, personnel, funding sources, methods, meetings, practises and so on, and that they can be examined collectively, as well as individually.

We do not argue that either approach (the individual or the sectoral) is superior to the other, simply that each offers an advantage for diverse sorts of analyses. We do feel that the sectoral approach has been less prominent and could be used more, and more effectively. We illustrate through some of the insights from the present survey, as well as work Dan Brockington has undertaken with development NGOs since the conservation NGO survey (Brockington 2014a).

This chapter proceeds as follows. First, we begin by reflecting on the nature of the relationship between anthropology and conservation organisations, detail some of the insights to be gained from studies of individual organisations, and consider what is missing. Then we report some of the findings of our sectoral approach, and outline the insights it offers.

Studying Conservation NGOs

Understanding conservation NGOs is a vital aspect of understanding conservation (cf. Adams 2004). These are large and complex organisations, which require care and effort to understand. They merit study in their own right because of the influence they have on the collective consciousness (in their brands and campaigns). They merit study for their reach and influence on conservation practises, particularly in poorer parts of the world where their resources are, relatively speaking, more influential. They merit study for the powerful ideas of nature and visions of conservation that they proselytise.

The distinguishing feature of the corpus of existing works that it tends to be based on studies of individual organisations—and to great effect and insight. There are some particularly remarkable doctoral theses demonstrating this. For example, Sarah Milne's examination an anonymised large international conservation organisation's work on payments for ecosystem services in the Cardamon mountains of Cambodia shows how the trial of a scheme to introduce payments for ecosystem services produced unintended and not always welcome consequences (Milne 2009). Milne found that social justice considerations were not well served because of the nature of rural politics in Cambodia and the organisation's unwillingness to challenge them on the ground.

Hassan Sachedina's thesis studied the African Wildlife Foundation (AWF) in Tanzania (Sachedina 2008). Hassan had worked part time as a fundraiser for the organisation and had wanted to continue that role after writing his thesis. His observations in the field disillusioned him however. He showed how the organisation had become successful as an organisation, growing in size and scope. Hassan also showed that this was not translating into effective conservation work on the ground, or indeed very much time in the villages. The AWF was most effective at being an organisation, but less effective at being a *conservation* organisation.

Both theses bring out common themes that are reported more broadly in the literature. One is that conservation NGOs have allowed themselves to be implicated too easily in heavy-handed state behaviour. Sunseri (2005) showed how alliances of local and international conservation NGOs are driving a new wave of evictions from forest reserves in Tanzania. Pearce (2005a, b) and others have written of the role of the African Parks Foundation in Ethiopia, which was implicated in the attempted clearance of the Guji from Nech-Sar National Park. Bonner (1993) described Prince Bernhard of the Netherlands channelling clandestine funds to the World Wide Fund for Nature (WWF) to support antipoaching work in Africa. Ellis reported the most unwelcome effects of that antipoaching activity on the elephant population (Ellis 1994).

The problematic social consequences of conservation actions are another abiding theme of studies of conservation NGOs. This was the Brockington's introduction to the topic when he found how successful the conservation promulgated by the George Adamson Wildlife Preservation Trust and sister organisations had been, but how questionable the social consequences of this intense conservation were (Brockington 2002). Much of the work being done there was brought together in a meeting of the 'Disobedient Knowledges' group in Washington DC, convened by Jim Igoe and Sian Sullivan; the meeting brought together scholars and activists to share a range of experiences of encounters, often bruising, with various conservation organisations. (Some of their experiences were collected in *Current Conservation*, 2010, Vol. 3, No. 3.)

In this corpus of work, which we have only begun to sketch in the preceding, two obvious points immediately arise that suggest clear advantages to approaching conservation NGOs as a sector. The first is that the conclusions and concerns of researchers exploring conservation NGOs echoes a slightly older issue raised by researchers working with development organisations (Edwards and Hulme 1992; Hulme and Edwards 1997; Bebbington et al. 2008). These authors also raised questions about the effectiveness of development organisations and their proximity to power. The detail of the similarity of these two bodies of work has been reviewed elsewhere (Brockington and Scholfield 2010c).

Nevertheless, the important point here is that there appears to be no good reason for this split between research on development NGOs and conservation NGOs. It was not even a conscious, deliberate disengagement based on sound reasons, or personal disagreement. On the contrary, neither group of researchers seems to be particularly conscious of the fact that they are ignoring each other. There is no good reason, to our mind, to separate conservation NGOs from development NGOs. This is partially because, conceptually, conservation is a form of development and partially, practically, because both groups of NGOs do very similar work on the ground in their real-world projects and lobbying, and they attend the same meetings to do so.

The second point is that it is only by treating conservation NGOs as part of a sector, and by comparing to other NGO sectors can this fact be realised and similarities between bodies of critique and literature be made. A sectoral approach enlarges the scope for comparative work. It would also then make it possible to explore how this sector might relate to other sectors and economic or social groups, such as the film industry or, indeed, the development NGO sector.

Note also that a common characteristic of current studies of conservation NGOs is that they have tended to be drawn to the work of the larger and powerful organisations (if not the largest, organisations—the ideas of which were still powerful locally). This has perhaps been the most well-articulated in Mac Chapin's famous 'Challenge to Conservationists', which appeared in the magazine *WorldWatch* in 2004 (Chapin 2004).

His targets were the big international organisations (BINGOs)—namely, the WWF, Conservation International (CI) and The Nature Conservancy (TNC)—that were, allegedly, affiliating with big business and industries and against indigenous and local inhabitants. Mark Dowie developed this theme in his writings, including the AWF in the list of problematic organisations (Dowie 2009). Others have contributed book-length critiques of single organisations (MacDonald 2008).

The concept of BINGOs usefully categories a group of NGOs, the size, reach and influence of which makes it helpful to treat as a group. Moreover, they behave as a group, as manifest, for example, in their associating to form the International Conservation Caucus, a powerful Washington-based lobbying group so thoroughly studied in Catherine Corson's work (2010). One collection of critiques questions the power of these large international NGOs over smaller conservation movements and state conservation departments. Rodriguez et al. (2007) warn against domination of weaker local partners by international organisations. Lees (2007) observes a proliferation of international interest in conservation in Fiji, but no increase in the effectiveness of conservation performance. Duffy (2006) observes that larger conservation NGOs wield considerable influence through the donor council in Madagascar, and they helped push for the establishment of more national parks in the country.

Understanding these politics are useful, but we have a nagging doubt as to the tendency to categorise conservation organisations dichotomously as the 'Big' and the 'rest'? Can we not breakdown the 'rest' into smaller groups, the function of which varies from others? As Table 3.1 shows, semi-seriously, on closer examination of the conservation NGOs operating in sub-Saharan Africa, a great variety of dissimilar groupings are possible.

Moreover, as we illustrate in the next section, some slightly more sensible categories based on financial power, geography of activity and position in funding hierarchies and networks also can be useful. A sectoral approach makes it possible to explore how these categories related to each other—funders, funded, moving of personnel and ideas. A sectoral approach is routine when it comes to, for example, trends in funding to certain groups of NGOs in Britain (Atkinson et al. 2012)—so why not apply a similar logic to conservation NGOs?

A sectoral approach, done well, also would consider the internal coherence of any categories identified, both with respect to the sector itself and to any groupings identified within it. The advantages of this sort of approach are clearly visible in George Holmes's work (Holmes 2010).

Table 3.1 Varieties of conservation NGOs

A. BLINGOs (BLImey that's a big NGO!) These are known elsewhere as BINGOs, but Bling is clearly more apt than mere Bing
B. FLAMINGOs (Fairly LArge MultI-million dollar NGOs) These are slightly smaller than the BINGOs but still spend millions of dollars a year
C. WANGOs (Wonderful Animal focussed NGOs) These organisations focus their African activities around the conservation of a particular species but can range in size and origin
D. PONGOs (wonderful Person fOcussed NGOs) These are conservation organisations devoted to saving wildlife but that have an appeal tied to the charismatic individuals who established and inspired them
E. HONGOs (Habitat fOcussed NGOs) These are NGOs that focus on various habitat
F. GNUNGOs (GeNUs focussed NGOs) These organisations focus on groups of animals, for instance, big cats, primates and so on
G. BONGOs (Bird fOcussed NGOs) These are similar to Genus-focussed NGOs, but they focus their conservation activities on birds. Better still would be SONGS ON BONGOS (SONGbird Or Non song Bird fOcussed NGOs)
H. SPANGOs (Single protected area NGOs) These are usually smaller organisations and focus all their attention on one specific area
I. DINGOs (Dabbling in Conservation NGOs) These are organisations linked to hunting clubs or tourism organisations, where perhaps conservation is the 'secondary' activity of the organisation
J. CONGOs (COmmunity-based NGOs) These are small conservation organisations that were set up by local groups
K. STINGOs (Student Inspired NGOs) These are organisations that have been set up by groups of friends, predominantly students, who had previously travelled to the area and decided they wanted to make a difference and set up a conservation organisation
L. MANGOs (MemoriAl NGOs) These are NGOs that have been set up in the name of deceased prominent conservation figure
M. GRINGOs (Good Research Involved NGOs) These are organisations that often have grown out of, or been established alongside, research projects
N. YOUCANGO (YOUth and Caring Adults NGOs) These groups provide paying volunteers for projects and journeys. It is not always clear how much of the revenues of these groups go to conservation projects and how much to the organisation
O. BANGOs (BriefcAse NGOs) These do not really exist outside of the briefcase carrying the fundraising proposals for their projects. A common problem, but it is difficult to name the organisations involved

(*continued*)

Table 3.1 (continued)

P. NGONGONGOs (Non-Governmentally Organised Networks and Groups of NGOs) These are large organisations that network African conservation organisations, although not necessarily implementing their own projects
Other categories we hope to populate in due course are: **OH NGO!** (the set of particularly silly NGOs); **BROWN NGOs** (the NGOs that will say anything to get approval); PINGOs (the cold hearted NGOs); **LINGO** (non-Anglophone NGOs); **SORRY I DON'T SPEAK THE LINGO** (Anglophone NGOs); **SONGs** (Francophone and musical); **WRONGs** (Francophone, but misguided); and **TANGOs** (pairs of NGOs moving in harmony, very rare)

Source: From Brockington (2011)

He examined the role of conservation elites and found, surprisingly, that transnational elites can be in remarkable conflict with national elites, particularly where those elites self-identify in opposition to international influence.

We submit, then, that even though individual studies of particular organisations in specific roles and tasks have provided a great deal of insight, there is much to be learned from treating NGOs as groups and networks and from examining them in toto, as well as on their own. Or, to put it in a slightly different way, that ethnographies of individual organisations could benefit from learning more about the context within which those organisations work. We are quite good at looking at national and political contexts, at general funding pressures and so on. Nonetheless, there is also a sectoral context—the group of organisations that conservation NGOs self-identify as belonging to—with tribulations that they share collectively, and with whom they are in competition with for funding and personnel. The sectoral context deserves more attention.

Let us outline briefly what was learned, when that sectoral approach was used in sub-Saharan Africa, to try to illustrate what we mean.

Basic Findings from the Sectoral Study

Some of the basic insights are geographical. If it is meaningful to group organisations together in a sector because of their similarities, then one can learn much from where similar activities are concentrated, where they are sparse, and from where their ideas and funding come.

Table 3.2 Location of head offices of conservation NGOs working in Africa

Country	*Number of head offices*	*Country*	*Number of head offices*
United States	65	Switzerland	3
United Kingdom	34	Uganda	2
South Africa	33	Belgium	1
Kenya	16	Burkina Faso	1
Namibia	11	Burundi	1
Tanzania	11	Denmark	1
Botswana	10	Djibouti	1
Madagascar	9	DRC	1
France	8	Egypt	1
Germany	7	Ethiopia	1
Netherlands	7	Gambia	1
Zimbabwe	7	Ghana	1
Malawi	6	Guinea-Bissau	1
Nigeria	6	Israel	1
Zambia	6	New Zealand	1
Cameroon	4	Portugal	1
Canada	3	Rwanda	1
Sierra Leone	3	Somalia	1
Australia	2	Sudan	1
Liberia	2	Tunisia	1
Norway	2	**Grand Total**	**278**

Source: Brockington and Scholfield (2010c)

With respect to conservation NGOs, we found that most of these organisations had, with the probable exception of organisations working in South Africa, headquarters based in the West or North (Table 3.2). We found that their spending and work were concentrated in East and South Africa, with relatively little in West Africa (Fig. 3.1). Other insights are historical; one can learn more about when a sector has been important, and how its prominence varies over time. Thus, we found that conservation NGOs working in sub-Saharan Africa had mushroomed during the 1990s to 2000s (Table 3.3).

Note that we are not claiming any revolutionary insights here. It is well known that NGOs generally blossomed under neoliberal dominance, which sought more efficient means than states for delivering services. Conservation NGOs are simply part of that boom. It is also hardly surprising that conservation NGOs are so clearly driven by Western desires, and presumably visions, for Africa, given the hold African landscapes and wildlife have in the Western imagination.

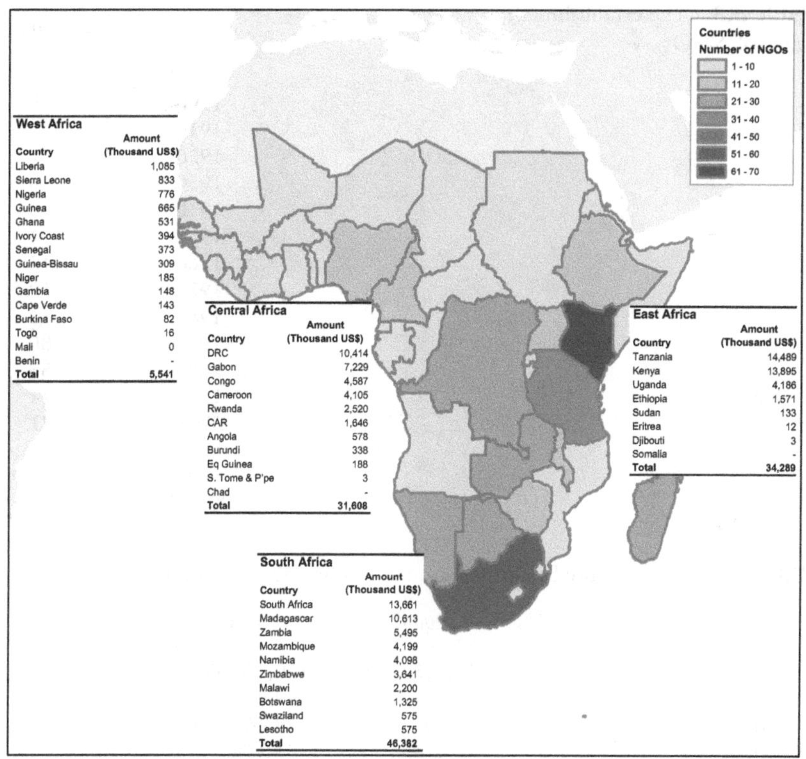

Fig. 3.1 The geographical distribution, by country, of conservation NGO activity and their spending in sub-Saharan Africa
Source: Brockington and Scholfield 2010b

Nonetheless, if the broad schema of these findings were already suspected, it is still useful to confirm them. It would help future historians to place studies of conservation before the 1990s into a proper context to know how active and numerous conservation NGOs were at that time. In addition, it is useful for comparative purposes to understand how removed West African conservation work is from that in South and East Africa.

Then there is the political economy of NGOs—how they create resources, for whom, and how they are distributed. It is, we submit, useful to know how much money is being mobilised by this sector, and how it compares with others (Table 3.4). It was, it appears, spending almost as

Table 3.3 The establishment dates of NGOs

Time period	*Count*
Prior to 1900	5
1900s	5
1910s	1
1920s	1
1930s	0
1940s	0
1950s	4
1960s	15
1970s	19
1980s	54
1990s	80
2000s	53
Unclear	44
Total	**281**

Source: Brockington and Scholfield (2010c)

Table 3.4 Expenditure by conservation NGOs in sub-Saharan Africa 2004–2006

	2004	*2005*	*2006*
Expenditure	113,723,444[a]	130,524,350	143,396,577
Expenditure, plus overheads	139,929,996	159,992,336	176,480,749
Overheads ($)	26,206,553	29,467,986	33,084,172
Overheads (%)	23%	23%	23%
Proportion of data in each year that has been estimated from other years	33.5%	0.4%	5.7%

Source: Brockington and Scholfield (2010b)

[a]All figures are express in USD as of 2006

much money as the hunting sector (USD200 million per year) when this survey was conducted (Lindsey et al. 2007).

It is useful also to know where those resources are concentrated, how unevenly the sector is structured, and who the major players are. We found that the conservation sector was, like other NGO sectors, remarkably uneven, with 10 major players accounting for most of the money spent (Table 3.5). Among them we found that the TNC was not nearly so important as elsewhere, and that there were two specific organisations for African conservation (i.e., Peace Parks and the AWF) that were particularly influential (Table 3.6).

Table 3.5 The observed and predicted structure of the conservation NGO sector in sub-Saharan Africa

Size class	*Range of expenditure including overhead (mil = million)*	*Counted NGOs*	*Average expenditure including overhead*	*Predicted NGOs*	*Predicted total expenditure including overhead*	*Predicted structure*
7	Over $40 mil	1	42,708,026	1	42,708,026	21%
6	$10 mil to $21 mil	4	15,559,663	4	62,238,653	31%
5	$4.2 mil to $6.2 mil	5	5,467,690	5	27,338,450	14%
4	$0.8 mil to $1.9 mil	10	1,351,520	18	24,327,360	12%
3	$0.3 mil to $0.72 mil	14	479,142	46	20,603,106	11%
2	$0.1 mil to $0.3 mil	26	200,090	90	18,008,100	9%
1	Up to $0.1 mil	27	54,927	102	5,712,408	3%
Total		**87**		**263**	**200,936,102**	

Source: Brockington and Scholfield (2010b)

Table 3.6 The 10 largest conservation NGOs in sub-Saharan Africa

Organisation name	*Average expenditure*	*Average expenditure including overheads*	*Countries where active*
World Wide Fund for Nature	35,212,994	42,708,026	44
Conservation International	17,264,283	20,247,980	9
Wildlife Conservation Society	15,585,563	17,321,231	19
African Wildlife Foundation	12,073,116	14,614,140	11
Peace Parks Foundation	8,392,335	10,055,302	9
Jane Goodall Institute	4,412,168	6,120,999	2
Fauna and Flora International	4,895,446	5,947,705	8
Frankfurt Zoological Society	4,837,535	5,895,838	7
African Parks Foundation	3,246,610	5,136,265	6
Dian Fossey Gorilla Fund	3,497,692	4,237,644	2

Source: Brockington and Scholfield (2010b)
Expenditure is in $US (2006)

We are making no grand claims about these data; they are all rather descriptive. Nonetheless, it does provide important context that other researchers will find useful in their examinations of other individual organisations, or for more comparative work.

Insights into the Functioning of Networks

Beyond the basic descriptive architecture and makeup of a sector, more detail is possible. In her PhD thesis Katherine Scholfield did a detailed examination of one conservation NGO sector (Scholfield 2013). She chose to research NGOs supporting mountain gorilla conservation in Rwanda and to explore how ideas about community interventions circulate. Essentially, she wanted to understand how NGOs—and other actors including states, communities and the private sector—interact, and what this means in terms of inclusion and exclusion of various people and groups and their ideas. She chose this sector because of the concentration and diversity of actors, particularly NGOs, involved and because of the opportunity it offered for researching a variety of interactions between them around a wide range of ideas and projects.

What she discovered was a complex mix of overlapping connections and disconnections, as Fig. 3.2 shows. The former was the result of project collaborations, government schemes and informal discussions both in and out of the office. The latter were sometimes strategic in nature, with NGOs choosing not to work together for reasons including organisational politics and competition for funding and publicity, but also because of the individual personalities involved.

Her work was able to illustrate how frequent and important the interchange of staff between organisations was (see Fig. 3.3). Indeed, in some sense, it does not seem right to speak of these as separate organisations—rather, there is a slowly evolving pool of people who simply coalesce into assorted configurations over time. This is fundamental to understanding the politics and interactions of NGOs; this is particularly the case, she found, wherever breaks or ruptures within that pool occurred. For given that they can function so closely, and be involved so much collaboration, divisions became especially significant.

Disconnections were also 'hidden' or 'accidental'. Exploring how the staff of these NGOs worked as a sector revealed how certain people and groups—particularly from local NGOs—might be disconnected from the wider network because they were not part of a given social circle, or did not have a significant level of perceived 'expertise'. This had an impact on whether and how they could discuss their ideas with other NGOs or with state representatives and whether they were invited to collaborate with other NGOs.

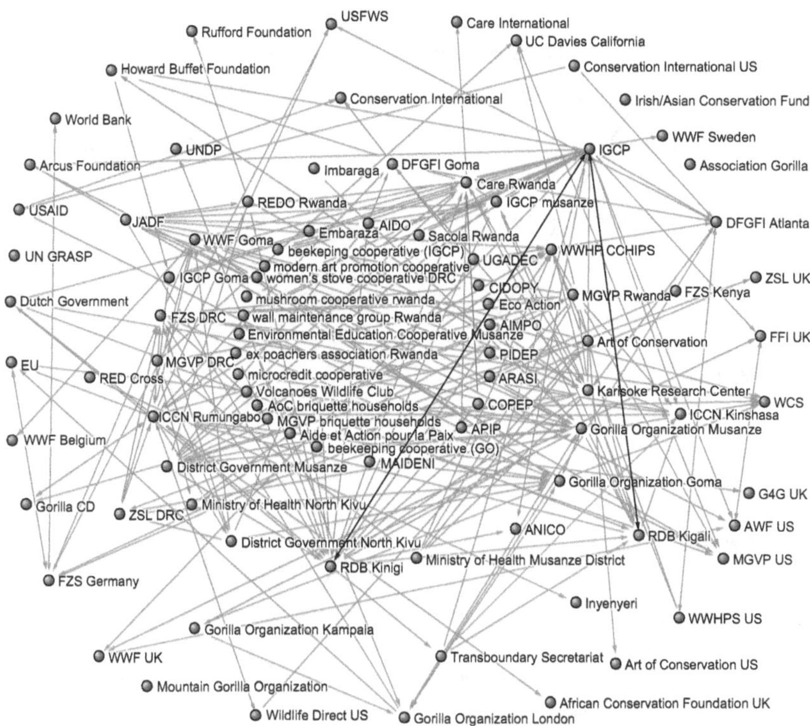

Fig. 3.2 Visualising the mountain gorilla conservation network
Source: Scholfield 2013

A further consequence to close collaboration and linkages between certain NGOs was how interdependent the ideas, plans and projects of them were. The NGO activities made little sense when viewed as the work of single organisations. For, in addition to personnel, they were also borrowing on and building on the latest ideas to flow through this particular conservation community. As Redford and Adams observed, conservation is faddish (Redford and Adams 2009). It seizes new ideas with enthusiasm, sometimes excessively so, without good reason. Besides, in the eagerness for briquettes, energy-efficient stoves, disease control and community-driven

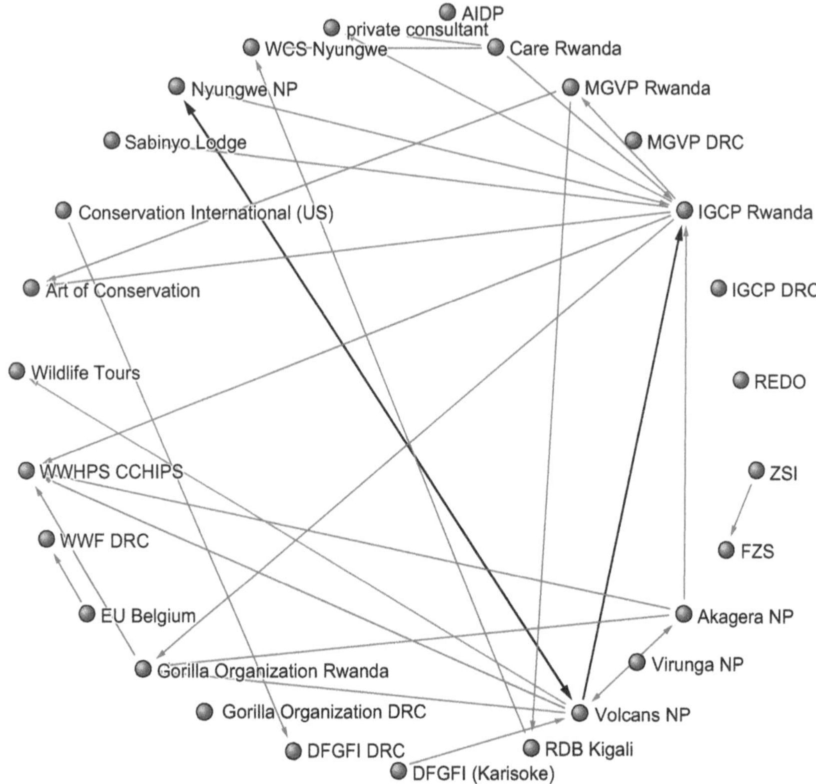

Fig. 3.3 Staff movement in Rwanda and the DRC
Source: Scholfield 2013

thinking sweeping this NGO community collectively, one could see these 'fads' at work; however, it was only with this sectoral perspective that this was possible. Likewise, it was only with this sectoral approach to understanding relationships between network actors that it was possible to then see whose ideas were marginalised, or excluded, from this process.

Researching the mountain gorilla conservation NGO network revealed what conservation looks like 'on the ground'; where the ground is not just a geographical location (although this forms part of it), but it is something that emerges as diverse types of interaction take place between people and organisations within particular spaces at specific times. It allowed for a clearer understanding of various types of interaction between NGOs, how these facilitate circulation of ideas about how things should be done, and what this means in terms of the inclusion and exclusion of certain people and groups.

Polemical Insights

A further advantage of the sectoral approach is that it allows groups of organisations to be examined collectively with respect to the goals that they share. This is perhaps most useful and interesting for applied anthropology, which seeks to engage and work with the organisations under study. The point here, however, is that one can do this from a perspective that the organisations themselves find interesting. It will be no less sensitive, but it still encourages engagement and interaction.

Consider, for example, NGO support for protected areas (PAs). Protected areas are an essential tool of the global conservation movement, providing key indicators and means of achieving conservation targets (Chape et al. 2005; Butchart et al. 2010); they are frequently financially supported by NGOs. Moreover, the costs of PAs often have been used to predict shortfalls in conservation expenditures, particularly shortfalls in the needs of conservation NGOs (James et al. 2001; Pimm et al. 2001; Balmford et al. 2003; Bruner et al. 2004; Moore et al. 2004).

Our survey of sub-Saharan African NGOs, their work and their expenditures, found that only 14% (by area) of all PAs on the continent, and 37% of the International Union for Conservation of Nature and Natural Resources (IUCN) category 1–4 PAs, received any form of support from conservation NGOs (Fig. 3.4). Nevertheless, significantly more support was theoretically possible. Following Moore et al.'s formula (2004; see Box 3.1), it would cost USD64 million to support *fully*, without any government expenditure, the annual running costs of 92% of the IUCN category 1–4 lands in sub-Saharan Africa (Brockington and Scholfield 2010b).

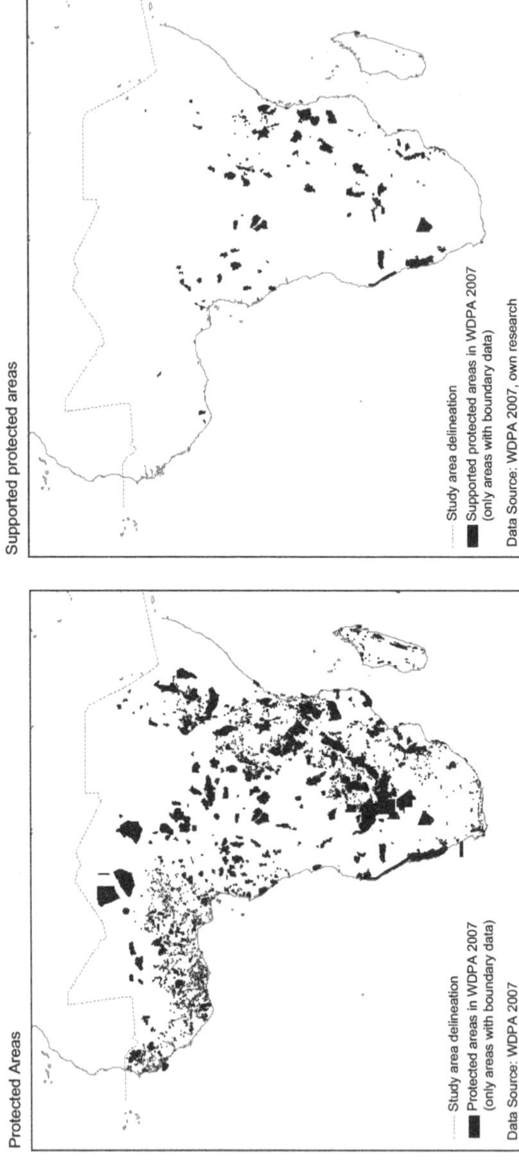

Fig. 3.4 The distribution of all protected areas, and those receiving some form of support from conservation NGOs in sub-Saharan Africa

Source: Brockington and Scholfield 2010b

Box 3.1 Equation Predicting Ongoing Management Costs of Protected Areas

$$\begin{aligned}\text{Log}\left(\text{Annual Cost; US\\$ km}^{-2}\text{year}^{-1}\right) = 1.765 - 0.299\log\left(\text{Area, km}^{-2}\right) \\ +1.014\log(\text{PPP}) \\ +0.53\log\left(\text{GNI US\\$ km}^{-2}\right) \\ -0.771 * \log\left(\text{Area, km}^{-2}\right) * \log(\text{PPP}).\end{aligned}$$

Annual observed conservation NGO expenditures in this region were just over USD140 million (the predicted expenditure was USD200 million). Thus, if conservation NGOs wanted to support a significantly larger area of protected land, they could. Indeed, collectively they could guard a far larger area more thoroughly than they currently do with less than half of their present revenues.

These figures are robust. The USD140 million observed expenditure was substantially less than the USD200 million predicted. In addition, the predicted expenditure required for PAs was lower than other estimates (Table 3.7). We are confident that the thought experiment we are suggesting here is not a figment of our data. Nonetheless, we need to be careful with insights such as these. Some observers insist that more money should be spent on PAs (Oates 1999; Terborgh 1999; Bruner et al. 2004), and they might conclude that these findings support their case. That conclusion would be premature.

There are good reasons *not* to spend NGO funds on PAs: protected areas may already be sufficiently supported by the state or relatively unimportant for achieving conservation goals. Alternatively, threats to them may be better tackled by actions beyond PA borders—for example, tackling illegal wildlife trade—rather than through direct support of the PAs themselves. Just because PAs *could* receive more assistance, does not mean that they *should*.

The difference between what NGOs could spend on PAs and what they actually do spend may represent a foolish use of resources, or a wise one. The point is that the relatively low cost of protecting existing protected areas, and (again relative) abundance of resources in conservation NGOs to provide them, had not been visible without a sectoral approach. Indeed,

Table 3.7 Comparison of the estimated costs of maintaining protected areas in cameroon based on three sources

Protected area name	*Cost per km² Moore et al.*	*Cost per km² Culvervel*	*Cost per km² Africa Resources Trust*	*Size km²*
Dja	134[a]	24	**153**	5,600
Faro	**196**	46	186	3,300
Campo—Ma'an	**233**	122	203	2,700
Bouba Ndjida	**273**	134	225	2,200
Benoue	**310**	183	247	1,800
Waza	**345**	207	255	1,700
Douala—Edéa	**366**	68	263	1,600
Korup	**364**	106	300	1,259
Banyang—Mbo	**542**	120	449	640
Kimbi	**2,825**	2,104	2,816	56
Kalamaloue	2,395	674	**3,396**	45
Lac Ossa	3,206	391	**3,762**	40
Mozogo—Gokoro	5,678	1,019	**9,646**	14
Total	**5,527,273**	2,090,528	5,072,323	

Source: Culvervel and ART are reported in Wilkie et al. (2001). Moore data are authors' calculations

Note: The highest estimates are in bold. Our prediction that it would be possible to support fully the expenses of more than 90% of strictly protected areas on just over USD60 million per year is based on supporting the largest and cheapest areas first. Smaller areas are more expensive to conserve. So, the fact that the formula produces low-priced costs for smaller protected areas compared to the ART makes no difference to the robustness of our prediction

[a]All figures are expressed in USD as of 2006

a sectoral approach is required to take it further. Despite the advances made by rational cost-effective conservation prioritising, comparative data are lacking; it is currently impossible to identify how the conservation NGO sector might identify the most effective strategies for collective interventions. A concentrated, more rigorous sectoral analysis of conservation spending and resultant benefits would make such an approach possible.

Calculation of the exact costs of conservation strategies requires a detailed breakdown of the convoluted paths between conservation intervention and conservation outcome. This is difficult because of the variety of activities undertaken; their inherent complexity; and varied transaction costs of collaborating with governments, international organisations, businesses, communities and other NGOs. There are also fundraising costs and administration expenses (23% of expenditures in Brockington and Scholfield's survey) that are not mentioned in high-profile estimates of conservation financial needs (e.g., Balmford et al. 2003). Still, progress is being made toward understanding how conservation interventions lead to outcomes (Kapos et al. 2008). Such analyses allow more sophisticated conservation bookkeeping that tightly links specific investment interventions to measurable conservation outcomes (e.g., maintaining a viable population of a threatened species, or the integrity of a unique ecosystem).

This final point may be the most important. Better coordinated activities might be more appealing to funders, but they would also greatly increase the capacity of the organisations themselves. Mercer's analysis of environmental philanthropy found most NGOs could only expand their current capacity by 20–30% (Mercer 2007: 114). Much greater expansion is required; however, simply adding more money to existing practises will not necessarily produce the desired conservation outcomes. There are well-documented instances of successful fundraising resulting in ineffective conservation (Dowie 2008; Sachedina 2010). A sectoral approach may facilitate a more effective scaling up of conservation NGO activity.

Analogies from Research into Development NGOs

One objection to a sectoral approach is that it creates categories that are meaningful to the analyst but not to those who actually are part of the sector. It creates categories that are heuristic devices to aid analysis but are of little relevance otherwise. Certainly, anyone who feels we have suggested that there should be a category of NGOs called BANGOs or NGONGONGOs (see Table 3.1) would be falling into that trap.

It is clearly important, however, to be aware of interlinkages and crossovers between organisations that make the basic unit of analysis (the NGO) rather unstable, as Scholfield's thesis shows. Or, indeed, there can be forms of cooperation and collaboration across organisations that make the behaviour of one more difficult to understand without examining the larger context. If, in fact, the object of our study is porous, fluid and merges with other objects then a step back to consider the larger context is needed, and again this requires a more sectoral approach.

It is easiest to illustrate this point by referring to some work beyond the conservation sector that Dan Brockington undertook to examine the role of celebrity in NGOs. The focus of this project was development NGOs, but the issue only can be fully comprehended by understanding what had happened to relationships between celebrity and NGOs as a sector.

Relationships between the celebrity industries and NGOs have been transformed during the last 15 years by a much more professional and systematic approach, particularly on behalf of the NGOs. This is visible in the rise of celebrity liaison officers, which now have been appointed by 75% of the largest NGOs in the United Kingdom. These officers meet once a month in London at the Celebrity Liaison Managers' Forum. The members have helped to shape the development of a niche aspect of the celebrity industries that cultivates and shapes interactions with NGOs.

It was possible for Brockington to write an ethnography, of sorts, that cut across the work of celebrity liaison officers in various organisations. They were facing similar pressures, frustrations and restrictions in their work. They had to struggle with similar misunderstandings as to the power and influence of celebrity among their colleagues—the same callousness and disregard as from so many players in the celebrity industries. They had to work around similar problems of demonstrating to the public the fit between celebrities and their organisations, and of cultivating relationships with the celebrities to strengthen that fit. They had to face the same problems of the unknown, and sometimes possibly slight, public response to the media events they initiated (Brockington 2014a, b).

The sectoral approach also made it easier to identify sectoral pressures and opportunities attendant to celebrity advocacy. This made it possible to observe how pervasive and far-reaching corporate influence was across NGOs on the rise of celebrity advocacy among them. It also was feasible to see how effective celebrity could be with respect to reaching and working with policy elites.

Such an 'ethnography' could not portray the detail of interactions in specific organisations. It could not afford the quality of insight that concentration on one organisation allowed. We are not, therefore, positing it as an alternative to studying individual organisations. Rather, it adds a necessary perspective and context that should make events in any one organisation more intelligible. Once again, this demonstrates that a sectoral approach makes it possible to see how NGOs are collectively responding to pressures and changing.

Concluding Remarks

In conclusion, analysing the expenditures and activities of conservation NGOs as a sector is important for several reasons. First, if one examines conservation NGOs on their own terms, it is easier to broaden the efficiency gains offered by systematic conservation planning. A better grasp of NGOs' collective current behaviour will facilitate the incorporation of the recommendations of systematic conservation planning into existing practise. It will permit more accurate estimates of conservation NGOs' financial needs. Existing estimates frequently are based on the maintenance costs of PAs (James et al. 1999; Bruner et al. 2004; Moore et al. 2004). Yet, these costs are clearly inappropriate because maintaining PAs accounts for little of what conservation NGOs actually do.

Conservationists insist that conservation NGOs need more money to halt biodiversity loss (Butchart et al. 2010). We still cannot, however, identify with any degree of accuracy where, or on which activities, extra resources should go. The high levels of transparency and accountability that NGOs can display individually are not apparent across the sector. More clarity about the patterns and consequences of existing conservation NGO expenditures would give appeals for additional funding greater credibility.

From a more critical perspective, a sectoral approach is necessary in part simply because of the actual practises that being an NGO entails. There is a great deal of cooperation in sharing ideas and personnel on the ground and over time. NGOs really cannot be considered as discrete entities. They are fluid, porous and malleable. In addition, even when there is not such dedicated collaboration, and when competition and separation are the order of the day, one still needs to understand that divorce properly to understand how the entities themselves behave. Competition over brand space and fundraising, for example, will require as much of a sectoral approach as understanding the International Gorilla Conservation Programme, or the IUCN.

There are of course disadvantages to this approach. It cannot replicate the detail and insights that good ethnography provides as to how decisions are made, and why particular organisations behave the way they do, from their internal workings. But then, there is not yet sufficient interest in sectoral approaches to dwell on these problems. These will become more significant if the balance of studies (between sectoral approaches and individual ethnographies) swings convincingly to the sectoral side; that is not yet the case. Therefore, this chapter's authors urge that more attention be paid to conservation NGOs as a sector in anthropology. There is much to be learned from greater scope, and comparisons, that this approach makes possible.

References

Adams, W.M. 2004. *Against Extinction: The Story of Conservation.* London: Earthscan.

Atkinson, A.B., P.G. Backus, J. Micklewright, C. Pharoah, and S. Schnepf. 2012. Charitable Giving for Overseas Development: UK Trends Over a Quarter Century. *Journal of the Royal Statistical Society, Series A* 175 (1): 167–190.

Balmford, A., K.J. Gaston, S. Blyth, A. James, and V. Kapos. 2003. Global Variation in Terrestrial Conservation Costs, Conservation Benefits, and Unmet Conservation Needs. *Proceedings of the National Academy of Sciences of the United States of America* 100 (3): 1046–1050.

Bebbington, A.J., S. Hickey, and D.C. Mitlin. 2008. *Can NGOs Make a Difference? The Challenge of Development Alternatives.* London: ZED.

Bonner, R. 1993. *At the Hand of Man. Peril and Hope for Africa's Wildlife.* London: Simon and Schuster.

Brockington, D. 2002. *Fortress Conservation. The Preservation of the Mkomazi Game Reserve, Tanzania.* Oxford: James Currey.

———. 2011. A Brief Guide to (Conservation) NGOs. *Current Conservation* 5 (1): 30–31.

———. 2014a. *Celebrity Advocacy and International Development.* London: Routledge.

———. 2014b. The Production and Construction of Celebrity Advocacy in International Development. *Third World Quarterly* 35 (1): 88–108.

Brockington, D., and K. Scholfield. 2010a. The Conservationist Mode of Production and Conservation NGOs in Sub-Saharan Africa. *Antipode* 42 (3): 551–575.

———. 2010b. Expenditure by Conservation Non-Governmental Organisations in Sub-Saharan Africa. *Conservation Letters* 3: 106–113.

———. 2010c. The Work of Conservation Organisations in Sub-Saharan Africa. *Journal of Modern African Studies* 48 (1): 1–33.

Bruner, A.G., R.E. Gullison, and A. Balmford. 2004. Financial Costs and Shortfalls of Managing and Expanding Protected-Area Systems in Developing Countries. *Bioscience* 54 (12): 1119–1126.

Butchart, S.H.M., M. Walpole, B. Collen, A. van Strien, J.P.W. Scharlemann, R.E.A. Almond, J.E.M. Baillie, B. Bomhard, C. Brown, J. Bruno, K.E. Carpenter, G.M. Carr, J. Chanson, A.M. Chenery, J. Csirke, N.C. Davidson, F. Dentener, M. Foster, A. Galli, J.N. Galloway, P. Genovesi, R.D. Gregory, M. Hockings, V. Kapos, J.F. Lamarque, F. Leverington, J. Loh, M.A. McGeoch, L. McRae, A. Minasyan, M.H. Morcillo, T.E.E. Oldfield, D. Pauly, S. Quader, C. Revenga, J.R. Sauer, B. Skolnik, D. Spear, D. Stanwell-Smith, S.N. Stuart, A. Symes, M. Tierney, T.D. Tyrrell, J.C. Vie, and R. Watson. 2010. Global Biodiversity: Indicators of Recent Declines. *Science* 328 (5982): 1164–1168.

Chape, S., J. Harrison, M. Spalding, and I. Lysenko. 2005. Measuring the Extent and Effectiveness of Protected Areas as an Indicator for Meeting Global Biodiversity Targets. *Philosophical Transactions of the Royal Society B* 360: 443–455.

Chapin, M. 2004. A Challenge to Conservationists. *World Watch Magazine*, November/December, 17–31.

Corson, C. 2010. Shifting Environmental Governance in a Neoliberal World: US AID for Conservation. *Antipode* 42 (3): 576–602.

Dowie, M. 2008. The Wrong Path to Conservation. *The Nation*, September 10.

———. 2009. *Conservation Refugees. The Hundred-Year Conflict between Global Conservation and Native Peoples.* Cambridge: MIT Press.

Duffy, R. 2006. Global Environmental Governance and the Politics of Ecotourism in Madagascar. *Journal of Ecotourism* 5 (1,2): 128–144.

Edwards, M., and D. Hulme. 1992. *Making a Difference: NGOs and Development in a Changing World.* London: Earthscan.

Ellis, S. 1994. Of Elephants and Men: Politics and Nature Conservation in South Africa. *Journal of Southern African Studies* 20 (1): 53–69.

Holmes, G. 2010. The Rich, the Powerful and the Endangered: Conservation Elites, Networks and the Dominican Republic. *Antipode* 42 (3): 624–646.

Hulme, D., and M. Edwards. 1997. *NGOs, States and Donors: Too Close for Comfort?* London: Macmillan.

James, A.N., K.J. Gaston, and A. Balmford. 1999. Balancing the Earth's Accounts. *Nature* 401: 323–324.

James, A., K.J. Gaston, and A. Balmford. 2001. Can We Afford to Conserve Biodiversity? *Bioscience* 51 (1): 43–52.

Kapos, V., A. Balmford, R. Aveling, P. Bubb, P. Carey, A. Entwistle, J. Hopkins, T. Mulliken, R. Safford, A. Stattersfield, M. Walpole, and A. Manica. 2008. Calibrating Conservation: New Tools for Measuring Success. *Conservation Letters* 1 (4): 155–164.

Lees, A. 2007. *Review and Analysis of Fiji's Conservation Sector.* Waitakere City: The Austral Foundation.

Lindsey, P.A., P.A. Roulet, and S.S. Romanach. 2007. Economic and Conservation Significance of the Trophy Hunting Industry in Sub-Saharan Africa. *Biological Conservation* 134: 455–469.

MacDonald, C. 2008. *Green, Inc. an Environmental Insider Reveals How a Good Cause Has Gone Bad.* Guidlford: The Lyons Press.

Mercer, B. 2007. *Green Philanthropy. Funding Charity Solutions to Environment Problems. A Guide to Donors and Funders.* London: New Philanthropy Capital.

Milne, S. 2009. *Global Ideas, Local Realities: The Political Ecology of Payments for Biodiversity Conservation Services in Cambodia.* University of Cambridge.

Moore, J., A. Balmford, T. Allnut, and N. Burgess. 2004. Integrating Costs into Conservation Planning Across Africa. *Biological Conservation* 117: 343–350.

Oates, J.F. 1999. *Myth and Reality in the Rain Forest: How Conservation Strategies Are Failing in West Africa*. Berkeley: University of California Press.
Pearce, F. 2005a. Big Game Losers. *New Scientist*, April 16, 21.
———. 2005b. Laird of Africa. *New Scientist*, August 13, 48–50.
Pimm, S.L., M. Ayres, A. Balmford, G. Branch, K. Brandon, T. Brooks, R. Bustamante, R. Costanza, R. Cowling, L.M. Curran, A. Dobson, S. Farber, G.A.B. da Fonseca, C. Gascon, R. Kitching, J. McNeely, T. Lovejoy, R.A. Mittermeier, N. Myers, J.A. Patz, B. Raffle, D. Rapport, P. Raven, C. Roberts, J.P. Rodriguez, A.B. Rylands, C. Tucker, C. Safina, C. Samper, M.L.J. Stiassny, J. Supriatna, D.H. Hall, and D. Wilcove. 2001. Environment—Can We Defy Nature's End? *Science* 293 (5538): 2207–2208.
Redford, K.H., and W.M. Adams. 2009. Payment for Ecosystem Services and the Challenge of Saving Nature. *Conservation Biology* 23 (4): 785–787.
Rodriguez, J.P., A.B. Taber, P. Daszak, R. Sukumar, C. Valladares-Padua, S. Padua, L.F. Aguirre, R.A. Medellin, M. Acosta, A.A. Aguirre, C. Bonacic, P. Bordino, J. Bruschini, D. Buchori, S. Gonzalez, T. Mathew, M. Mendez, L. Mugica, L.F. Pacheco, A.P. Dobson, and M. Pearl. 2007. Environment—Globalization of Conservation: A View from the South. *Science* 317 (5839): 755–756.
Sachedina, H. 2008. *Wildlife Are Our Oil: Conservation, Livelihoods and NGOs in the Tarangire Ecosystem, Tanzania*. School of Geography, University of Oxford Research Archive.
———. 2010. Disconnected Nature: The Scaling Up of African Wildlife Foundation and Its Impacts on Biodiversity Conservation and Local Livelihoods. *Antipode* 42 (3): 603–623.
Scholfield, K. 2013. *Transnational (Dis)Connections: Mountain Gorilla Conservation in Rwanda and the DRC*. D.Ph. Thesis, University of Manchester.
Scholfield, K. and D. Brockington. 2008. *The Work of Non-Governmental Organisations in African Wildlife Conservation: A Preliminary Analysis*. http://www.sed.manchester.ac.uk/idpm/research/africanwildlife/. Accessed 13 July 2009.
Sunseri, T. 2005. 'Something Else to Burn': Forest Squatters, Conservationists, and the State in Modern Tanzania. *Journal of Modern African Studies* 43 (4): 609–640.
Terborgh, J. 1999. *Requiem for Nature*. Washington, DC: Island Press.
Wilkie, D.S., J.F. Carpenter, and Q. Zhang. 2001. The Under-Financing of Protected Areas in the Congo Basin: So Many Parks and So Little Willingness-to-Pay. *Biodiversity and Conservation* 10: 691–709.

CHAPTER 4

Business, Biodiversity and New 'Fields' of Conservation: The World Conservation Congress and the Renegotiation of Organisational Order

Kenneth Iain MacDonald

INTRODUCTION

On 8 October 2008, three days in to the ten-day long World Conservation Congress (WCC), the blog *Green Inc.*, hosted on the website of the *New York Times*, ran this lead: 'This week at the World Conservation Congress in Barcelona, James Kantner takes note of a singular sentiment running through the participants: Conservation is failing because it has not embraced the fundamental tenets of business management'. This statement is remarkable on several accounts. First, it is inaccurate. Having tracked the issue for more than a decade, it is clear that what the International Union for the Conservation of Nature (IUCN) and other conservation organisations label 'private sector engagement' is an extremely contentious subject within these organisations. Indeed, debate during the Members' Assembly

K.I. MacDonald (✉)
Department of Geography and Center for Diaspora and Transnational Studies, University of Toronto, 1265 Military Trail, Toronto, ON M1C 1A4, Canada

P.B. Larsen, D. Brockington (eds.), *The Anthropology of Conservation NGOs*, Palgrave Studies in Anthropology of Sustainability, DOI 10.1007/978-3-319-60579-1_4

component of the WCC and the 'contact groups' established in relation to several motions from members seeking to limit the IUCN's 'engagement with the private sector' exposed a definite absence of 'singularity'.[1]

More important, Kantner's piece is notable because it serves to communicate two of the primary messages that the conveners of the conference—the IUCN Secretariat—sought to put into public circulation: (1) That it is time for the conservation movement to seek out and work with 'new types of partners' (i.e., code for deepening interconnections with the private sector), and (2) that the IUCN membership is unified in this pursuit.[2] That a journalist could identify these markedly political messages early on in the WCC meeting is not particularly surprising.[3] Indeed, the purpose of this chapter is to analyse the role of the WCC as a vehicle through which certain actors negotiate a new organisational order, reproducing the pseudo-social networks that are beginning to resemble, what Peter Haas (1992) and others have called, an 'epistemic community'; such is being directed through interaction among a set of common individuals who physically and ideologically migrate across the once well-defined boundaries separating governmental agencies, nongovernmental organisations (NGOs) and the private sector.

A primary point of the chapter is that we can attend to meetings like the WCC as instruments that facilitate and reveal the blurring of these boundaries, and as sites where the personal associations and ideological work necessary for the renegotiation of organisational order are acted out. To do so, however, we need effective analytic devices. Here, this author builds on recent approaches to analysing the performative aspects of governance (e.g., Hajer and Versteeg 2005; Hajer 2006) and suggests three such devices: (1) structure, (2) orchestration and (3) spectacle. These are used to analyse the WCC as an integral mechanism for achieving a renegotiated 'order' of conservation with 'private sector engagement' as a core operational practise. This chapter builds on ongoing work that situates the expanding integration of the 'private sector' and conservation NGOs in relation to neoliberal modes of environmental governance put in place over the past 20 years. Then, it highlights the need to study the interconnected web of related meetings to gain insight into the structuring of institutions and organisations engaged in the governance and the practise of biodiversity conservation (MacDonald 2003, 2008).

'Nature Is Our Business': Conservation, 'Private Sector Engagement' and the Culture of Meetings

To understand the rise in the association between private sector actors and biodiversity conservation organisations and institutions, it is necessary to view modernist biodiversity conservation as an organised political project.[4] There are two important dimensions to this claim. First is the recognition that the organisational dimensions of conservation exist as coordinated agreement and action among a variety of actors that take shape within radically asymmetrical power relationships. Second, the practical expression of that coordination exists as organised social groups—conservation organisations—that have emerged out of specific historical contexts.

Both aspects of 'organisation' in this context imply the promotion of certain ideological perspectives that are worked out through processes of coordinated agreement and implemented through the actions of conservation organisations. These are by no means exclusive processes. Indeed, the actions of conservation organisations are directed through the ideological configurations brought to bear on them by the coordinated agreement of relevant actors, or what have become known as 'stakeholders' and, in some cases, 'partners'. Such individuals, organisations and governments help to define what 'biodiversity conservation' is and provide the conceptual tools, material resources and political permission needed to do the work of conservation. This is not a new process.

Historically, conservation has been configured in relation, for example, to political projects of colonialism, nationalism and science. The contemporary emergence of business, or the private sector, as a major actor in shaping contemporary biodiversity conservation is, in many ways, a reflection of the coordinating action of global capitalism; its affiliated transnational capitalist class; and the need to redefine conservation in ways that accommodate, rather than challenge, the dominant ideological and material interests that underlie these broad political projects. As 'the environment' has turned out to be a transnational ideological formation with the popular capacity to challenge the interests of capital accumulation, it has become increasingly important for these actors to contain the capacity of emerging institutions and organisations to delimit and to constrain access to the natural foundation of capitalism.

Just 20 years ago, the reputations of conservation organisations would have been seriously compromised if knowledge of these 'engagements' had become public.[5] Today, however, the logos of conservation organisations, such as IUCN, Conservation International (CI), the Worldwide Fund for Nature (WWF) and Birdlife International, appear side-by-side on the covers of reports or in advertisements, with those of extractive industries (e.g., Shell, Rio Tinto, Total and Holcim). Appearing in dissimilar vernacular guises—sponsorships, agreements, partnerships—these private sector engagements have become almost ubiquitous among conservation organisations, and it is increasingly common to see the senior executives of larger organisations sharing convention stages and travelling with CEOs of companies historically engaged in environmentally and socially destructive activities.

Conservation organisations ground their justification for these relationships in their mandate to influence society in ways that result in environmental sustainability. More nuanced perspectives point to the mobilising effect of neoliberalism in establishing 'public–private partnerships' as fiscal cutbacks opened the door for conservation NGOs to expand project-based activities; in addition, it encouraged them to turn to private sector actors for material and political resources (Poncelot 2004). They also suggest that the effect of this 'engagement' is far from unilateral, that it has a significant effect in restructuring the ideological and organisational orientation of conservation organisations, and that it can be seen as an outcome of business organisations seeking to control an external institutional environment (Levy 2005; MacDonald 2010). Evidence of this shift can be seen more clearly in the rise of organisational units inside conservation organisations dedicated to establishing, fostering and managing collaborative relationships with private sector interests; social and environmental responsibility programmes established within corporate sectors; and incentive programs developed at state and suprastate levels to promote the establishment of such relationships (MacDonald 2008). These organisational shifts have led to an increase in conservation programming focussed on market-based conservation incentives, many of them grounded in dubious equations between ecological modernisation and sustainable development (Bruno and Karliner 2002; Elbers 2004; Frynas 2005).

The speed with which these relationships have travelled from the backrooms to the public stage at major conservation events has been remarkable, but it has not occurred without resistance. As recently as 2003, a plenary session at the World Parks Congress in Durban, South Africa, which included representatives from Shell and the International Council on Mining and Minerals (ICMM),[6] caused vocal protests from the participants, leading the IUCN's Director General, Achim Steiner, to have the audience microphones shut off, presumably in order to quell the protest.[7] Supporters of 'Business and Biodiversity Initiatives' also confronted significant vocal opposition at the 2004 World Conservation Congress in Bangkok, although this time it was tempered by the willingness of some IUCN members to allow the Secretariat to cautiously engage in discussions with the private sector. Recognising that the Secretariat was intensifying 'engagement' with the private sector, member-supported resolutions passed at the 2004 WCC were worded to ensure that this be done with due regard for the principles of transparency and accountability to IUCN members.

Four years later at the WCC in Barcelona, the public presence of the private sector had become much more apparent. But then again, the voice of protest, at least during the World Conservation Forum, was much less apparent.[8] This silence is explained in part through 'promotion' and 'presence'. In the years between 2004 and 2008, the promotion of 'partnerships' became much more prevalent in meetings, in funding programs and in joint publications, and this established a physical presence through mechanisms such as secondments of staff between the private sector and conservation organisations (e.g., IUCN).[9] Following the lead of government agencies and programme-defining organisations, such as the United Nations (UN), increasing numbers of NGOs had embraced models of public–private partnerships in pursuit of their policy and project goals.[10] Even though there is continued organisational resistance to that presence among the IUCN membership, the leadership of its Secretariat, and certain other actors within the organisation, have assumed the obligation not only to defend but also to extend that presence by attempting to legitimate it. There are various ways in which this legitimation occurs. In this chapter, however, this author makes the case that one primary mechanism is through, what Van Vree (1999) has called, the 'culture of meetings'.

Why Meetings?

To a remarkable extent, and despite their centrality in 'professional' life, 'field-configuring' events in conservation—the conventions and congresses where people gather to debate and formalise the direction of conservation policy and practise—have escaped the ethnographic lens (but see Poncelet 1990; MacAloon 1992; Little 1995; Dahlén 1997; Reed 2001). Certainly, emerging concerns over international environmental governance and the application of regime theory and analysis over the past 20 years has highlighted the importance of international conventions. Nonetheless, to a remarkable degree, this work is empirically distant from the physical sites where actual negotiation occurs. In addition, although ethnographic work that might act as a guide has begun (e.g., Nader 1972; Rosen 1991; Harper 1998; Fox 1998; Riles 2000; Markowitz 2001; Mosse 2001; Lewis 2003), what this chapter's author calls the 'organisation of conservation' remains understudied (MacDonald 2010). What little work that does exist is marked by disciplinary and scalar divides.

Typically, much of the work at the 'community' level is undertaken by anthropologists and geographers, while analyses of how conservation is structured through international institutions is seen to be the ambit of political science and sociology; this means that those who work in 'the village' are rarely the same as those who study the 'international regimes' (Lahsen 2007). This is a worrying phenomenon given the increasing capacity of those regimes to shape the ideological and practical work of conservation organisations; draw together actors that seek to structure those regimes; and, ultimately, to shape ecological dynamics.

The failure of ethnographers to attend to these sites is doubly troubling because of the abstract quality of much work in regime theory and transnational environmental governance. Rather than attending to questions of how the 'process' of negotiation or interaction expresses a specific cultural–political history and shapes the outcome of conventions, agreements or organisational mandates, regime work is primarily concerned with the 'outcome' and is empirically grounded in textual analysis and representations of interaction instead of direct observation of those interactions. As a result, much work on governance regimes is inadequately contextualised. It relies on the disassembling of text and individual memory and helps to aggregate institutional facts in centres of accumulation that, much like laboratories, are configured around the dominant theoretical perspectives of researchers. Nonetheless, this work appears blind to its role in the creation of institutional facts (Latour and Woolgar 1986).

Within conventional regime analysis, what counts as data, and how they are sanctioned, is grounded in the understanding and subsequent analytic treatment and representation of institutions or organisations as objects. This is far from accurate however; institutions and organisations are the dynamic result of intentional interaction among a group of actors with diverse interests. They are dynamic social constructions. If we seek to understand the effect of institutional or organisational actions, we need to pay attention to the events that not only facilitate interaction but also serve as instruments that can be used to direct organisational and institutional structure (Pulman-Jones 2001).[11]

Despite the surprising degree to which regime analysis is divorced from empirical settings (e.g., meetings of conventions or organisations related to environmental governance), internal critiques have slowly developed. Litfin (1994: 177), for example, in advocating what she calls a 'reflectivist approach' grounded in discourse analysis, notes that dominant approaches to the study of environmental governance regimes 'fail to grasp the nonmaterial nature of knowledge-based power; nor do they dig beneath the surface to explore the process of interest formation'. She finds similar flaws in work on epistemic communities, citing a serious underestimations of the degree to which the 'scientific facts' introduced to governance mechanisms are not divorced from political power but rationalised and reinforce existing political conflicts. Liftin's analysis leads her to conclude that all the dominant theoretical approaches to international regime analysis neglect the role of intersubjective understandings as the basis for international cooperation.[12]

Vogler (2003: 27) amplifies Lifton's points by deftly pointing out that 'while regimes are social and intersubjective phenomena their contemporary analysis is distinctly positivist—setting up an apparent contradiction between ontology and epistemology'.[13] In their critiques Vogler and others question the analytical capacity of regime analysis—given its intergovernmental orientation and positivist bent—to assess and explain the emergence of forms of transnational environmental governance in which a diversity of actors, with very real material interests in the potential constraining or facilitating effect of policy and international environmental law, struggle to shape the policy process. Although some work on regime analysis seems to have recognised the importance of the subjective quality of institutions and institutional facts, they have not grappled with what this means in terms of appropriate sites and scales of observation, methodological approaches, the constitution of 'data', and how that data is sanctioned.

Regime analysis suffers from a failure to recognise that institutional facts do not exist a priori but are created through political contest, and their separation from the social context in which they were created is only the final stage in their creation (Latour and Woolgar 1986). The point here is that, in failing to attend to the social relations and controversies involved in the reproduction of institutions and organisations, regime analysis plays a role in manufacturing institutional facts by effectively separating them from the political challenges in which they emerge. Not being present at sites where political disputes are acted out means that the trace of associations that define these political contests are not seen.

MacKenzie (2009: 33) and others have noted that this failure to pay due attention to these 'details' or 'technicalities' stems from a scale bias that reflects a tendency to divide 'social' phenomena into 'small, "micro" phenomena' (e.g., interpersonal interactions) and 'big, "macro" phenomena' (e.g., the international system of states), and to think only of the macro as political; however, the work of MacKenzie (2009) and Latour (2007) among others has also effectively pointed out the dynamics of scale and how 'micro' and 'macro' matters can switch positions over a very short span of time.

For regime analysis, then, institutions and organisations are objects of study and not sites of study (e.g., Hutton and Dickson 2000). Accordingly, they do not pay close analytic attention to the events and meetings that give them substance. The scale of analysis, mechanisms of abstraction and the failure to interact with and observe real actors engaged in decision making and the staking out of positions misses the ways in which such positions take shape across time and space, and the formation of associations that contribute to those positions.[14]

Attending to these events is important because the emergence of transnational environmental governance, the consequent threat of regulation, and the accordant possibility of subordinating some interests in 'the environment' has drawn previously separated actors together into spaces in which claims over 'nature', and the ideological and material struggles that lie underneath those claims, not only become unavoidable but also more readily visible and subject to scrutiny (Latour 2004). Within (and beyond) these spaces, actors intentionally seek to give substance to the institutions and organisations engaged in environmental governance in ways that express that interest. These locales, then, although not necessarily privileged, become important sites from which to compile accounts of these interests; they are places where the stakes of actors are articulated, where actions and associations formed in relation to those stakes become visible, where dissension within and between groups becomes apparent, and where contestation over the shaping of conservation policy and practise becomes clear.

These will not be found in the transcripts of official sessions, or filtered through the meta-language of the regime analyst, but by paying attention to the asides, by observing associations that are not subject to the official record, by noticing the encounters that were not witnessed by interview subjects, and by noting the tone of voice that cannot be 'heard' in the final report. Through being present at the site, analysts can record the process of knowledge being translated and observe how it gains traction in relation to specific interests. They witness meaning as it is being made, challenged, transformed and translated. In addition, they are exposed to the agency of those involved in that process. They are not restricted to the archives, whether textual or human, and are in fact better placed to evaluate the cogency of any documents.

The point here is that efforts to understand how transnational environmental governance is structured in relation to certain interests need to attend and attend to the sites in which diverse actors come together to debate and produce the structure of the institutions and organisations that assume, or are mandated with, the responsibility for that governance. It is at these sites where the ideological and material struggles that underlie environmental politics are revealed. Besides, we also need to be alert to the ways in which the sites (e.g., meetings and events) are malleable social phenomenon rather than absolute ostensible mechanisms. Events, such as the WCC meetings, are not simply important for *what* they accomplish, but for *how* they accomplish it. As one of the few scholars to focus on the history and evolution of 'the meeting', Van Vree (1999) described how meetings have come to provide an organisational mechanism and symbolic device to effectively regulate social behaviour and emotional life.[15] One of his key insights is that ongoing meetings help to produce enough permanence and cohesion to create an organisational culture in which norms of individual and collective behaviour become easily known, learned and circulated, and that meetings and the norms they invoke work to minimise the instability and uncertainty of random engagements, relegating such encounters to the category of awkward incidents.[16] Reed (2001: 133) commenting on Van Vree's work adds:

> [T]he enhanced power of the transnational intergovernmental organizations [from the mid-twentieth century onward] produced a meeting culture around the behavioural norms of a professional and business elite. These norms focus around more informal nodes of meeting behaviour where self-control and self-management are seen to be preconditions for the negotiation and renegotiation of the organizational order.[17]

Without invoking Foucault, Van Vree (1999: 331) effectively describes the disciplining quality of meetings, as learning meeting rules became an important component of societal success that instilled the internalisation of appropriate meeting behaviour, which he describes as a relatively 'precise, constant and smooth self-constraint of expressions of affect and emotion'.

His point here, one that I build on later, is that the 'culture of meetings' serves to communicate modes of acceptable behaviour so that new ideas, circulated by dominant interests, can gain more purchase with minimal opposition. This insight helps to situate meetings (e.g., the WCC) and their role in what Reed (2001) calls the renegotiation of organisational order; for as new actors (e.g., the 'private sector') have, for a variety of reasons, become central to the organisational interests of conservation, it has become necessary to reconfigure the organisational terrain to accommodate those interests. In the case of the increasing presence of controversial private sector actors in the day-to-day work of conservation organisations such as IUCN, we can understand the minimisation and containment of dissent; this is possible by paying attention to the ways in which meetings and events rely on the disciplined docility that characterises 'meeting behaviour' and can be structured and orchestrated in ways that effectively serve to legitimate the renegotiation of organisational order.

Meetings as a New 'Field'

Although the application of ethnographic methods to conservation is not new, the study of conservation events that this author, and others, advocate in this book means reconsidering what we have usually referred to as 'the field'. To study conservation as a dynamic political project from an ethnographic perspective means bringing spatially and temporally distant conferences and meetings within the purview of that 'field'. This is strange ground for ethnographic practise, which has historically sought field sites that, while dynamic, tend to endure in time and space, can be revisited, and made subject to renewed interpretation. Nevertheless, even though events like the WCC are more limited in time and space than conventional field sites, it is an empirical and methodological mistake to assume that they are isolated and temporary events. When we think of meetings as locales in a 'translocal' field—as nodes linked together across space and time by associations, common interests, long-term objectives, long-term agendas and statutes—the events take on a temporal and spatial durability that is obscured by the physical space of the encounter. They extend and repeat across time and space, bringing the same and different actors together at regular intervals, and their temporal durability creates the ability to strategise and plan for the events in ways that bring specific actors and interests together, introduces new actors to the assemblage—as has happened with 'private sector' participants in the network of conservation meetings—and structures their encounters.

As an example, preparing for one of the many meetings related to the Convention on Biological Diversity (CBD) consumes the time and effort of large numbers of people inside government ministries, NGOs, indigenous communities, UN agencies, universities and trade associations, among other groups. These people may only occupy a common space for 10 days every two years, but planning for 'the meeting', actions based on 'the meeting', and negotiations for structuring the next meeting, go on and push at the edges of what we typically think of as the temporal and spatial boundaries that contain the social encounter of 'the meeting'. There are seasons and cycles of activity that draw people together in ways that give life to the meeting but are not constrained to the 'time frame' or space of the event itself.[18]

'Opening' and 'Closing' ceremonies are in fact no such thing; they neither 'open' nor 'close' the meeting for anyone directly associated with it. They are symbolic, an adherence to ritual—those 'civilised norms' that travel through history in the guise of protocol. As anyone who has participated in meetings knows, 'the meeting' is constantly present: preparing presentations, following-up on delegated responsibilities, taking on new work as a function of 'the meeting', establishing networks, generating new facts—all of these tasks challenge the bounded quality of 'the meeting'.

The point here is that durable connections that stretch across time and space effectively challenge the idea of a meeting as an encounter that is isolated in time and space. This is made clear when the meeting is seen as a mechanism through which ideological perspectives can be circulated, gain traction and begin to structure both policy and the material practise of biodiversity conservation. The WCC, as an example, is an event that lends a consistent, if periodic, ability to observe the ideological and material struggles central to defining and operationalising the economic dimensions of biodiversity conservation—'making the business case for biodiversity and the biodiversity case for business', as IUCN's Chief Economist put it. Nonetheless, when it is combined with other conservation meetings and considered in the context of broader research—in my case, a project devoted to an ethnographically informed understanding of the effects of neoliberalism on the ideological and material orientation of conservation organisations—it becomes a node in a translocal field site. It is one constituted not simply by individual sites but by the relationships between mobile actors who, as they work *in advance* to plan meetings, come face-to-face *at* meetings, communicate *between* meetings, and pursue common objectives *through* the mechanism of meetings, and lends shape and coherence to that field site. The 'field', in this case, is not neatly delineated and contiguous as, say, a 'village'—in many ways it mimics the rhizomic structure of transnational space—but it is just as coherent.

In my work I, in part, have been tracking the 'associations'[19] of people who serve as business and biodiversity consultants: (1) their patterns of collaboration; (2) the events and objects that organise their work; and (3) the partnerships that evolve between particular individuals in relation to specific projects. The latter is particularly between those who circulate through organisations as consultants and those who I call the anchors—the permanent organisational employees who share the ideological perspectives and facilitate the access of consultants to organisations and project-based financing. Also in this work, it has become obvious that meetings play a key role in reproducing and solidifying these relationships. The WCC, then, is one in a group of interconnected meetings being 'followed'—a node in a network of events in which people, ideas and objects can be tracked to understand the ways in which they are orchestrated and configured and to identify the means through which such associations become far-reaching and durable. Of course, this is an element of the work that goes into producing the net*work*. As Hannerz (2003: 206) points out in his discussion of multi-sited fieldwork:

> [S]ites are connected with one another in such ways that the relationships between them are as important for this formulation as the relationships within them; the fields are not some mere collection of local units. One must establish the translocal linkages, and the interconnections between those and whatever local bundles of relationships which are also part of the study.

Hannerz's point is that 'the field' is constituted by the relationships that are being followed. As we try to grapple with applying ethnographic methods to ever more mobile people and ideas that ideologically and materially shape contemporary biodiversity conservation, it is important to focus on organisational actors (e.g., the IUCN) and their position within regimes that constitute transnational environmental governance, and on those seemingly temporary sites and events (e.g., the WCC). It is necessary, however, to situate those events within a process of longitudinal and multi-sited fieldwork that treats policy-setting venues as one node through which to understand the formation and extension of associations, the circulation of certain ideological perspectives through those associations, and the interlinked material interests that such networks and venues support.

This focus pushes research on the regulation and governance of biodiversity conservation to move beyond the abstract and to ground insights in direct observations of interaction and mobilisation as they occur.

Structure, Orchestration and Spectacle: Renegotiating the Organisational Order of the IUCN

Approaching the Meeting

Studying meetings means something more than simply attending sessions and analysing the discursive content and intent of the sessions. It means situating that content within an analytic frame that better explains the composition of the meeting as a political phenomenon designed in relation to the, often contested, ideological and material intent of its conveners. It frequently is difficult to solicit meaningful statements of intent from conveners, which is what makes an ethnographic approach to meetings particularly valuable; this is because it provides the capacity to align observations of the material structure of the meeting, and the symbolism deployed, with participants' statements in a way that yields insight into processes of orchestration.

In what follows I build on observations as a participant-observer at the WCC to develop a framework for analysis that provides insight into the ways the event can be understood as an instrument to better align organised conservation with global capitalism; concepts of ecological modernisation that underpin emerging market devices meant to allocate biodiversity, conceived as 'ecosystem services'; and the direct integration of the private sector in new forms of environmental management. These shifts are not new.

An initial tentative engagement of the IUCN Secretariat with concepts of ecological modernisation can be traced to the early 1980s but only became prominent in the early 2000s (MacDonald 2008). Dissent within the organisation became clearly apparent as the Secretariat began to establish mechanisms, such as the Business and Biodiversity Initiative,[20] to integrate private sector actors within the organisation. Even though there was dissent among the Secretariat staff, this was easily stifled through the alienation of the dissenting personnel. Nevertheless, dissent within the membership quickly became a flashpoint at meetings and took the senior staff of the IUCN somewhat by surprise. By 2003 the IUCN had established staff secondment agreements with extractive industries such as Shell. One effect of this became apparent during the 2003 World Parks Congress, held in Durban, South Africa.[21]

During the Congress, IUCN Secretariat staff, seconded to Shell, gave PowerPoint presentations bearing the Shell logo, encouraging protected area managers to establish partnerships with extractive industries; they emphasised the resources that the private sector could bring to protected area (PA) management, including skills in Geographic Information Systems, transportation capacity and 'human resources'. The goal was clearly to encourage fiscally beleaguered PA managers to see the material benefits of engaging in access agreements with the extractive sector. Although these sessions occurred in small backrooms, a larger plenary panel featured executives from extractive industries including Shell, British Petroleum (BP) and the International Council on Mining and Metals (ICMM).[22] Following this session audience microphones were ordered shut off by IUCN staff as the discussion period began to raise some troubling questions about the environmental impacts of these actors and potential damage they could cause to the IUCN's credibility as an environmental organisation. In subsequent sessions with the extractive industries, audiences were informed that there was no time for questions.

The issue of what the IUCN labels 'private sector engagement' also was contentious at the 2004 World Conservation Congress in Bangkok. Two member-supported resolutions instructed the Director General on how to proceed with this engagement and required that it be done transparently and in consultation with the IUCN membership. In the intervening four years at the 2004 and 2008 WCCs, I had begun to analyse the intensified process of 'private sector engagement' that had acquired a more concentrated focus within the IUCN. It was clear from this work that while there was an explicit commitment of the Secretariat, with the support of the Director General, for more intensified relationships with the private sector, much of the membership disputed the Secretariat's mandate to pursue these relationships. The 2008 WCC marked an opportunity not only to 'track' the presence of business but also to follow the related associations and the tensions within and between the various components of the IUCN over 'engagement with the private sector' and to observe how this tension would be managed in the context of the Congress.

The origins of this work clearly were exploratory but through a combination of participant observation, discussions with key informants, and interviews with participants, it became apparent that the WCC served as an instrument in the negotiation of a new organisational order. It was to be one that not only sought to legitimate the presence of new conservation innovations and actors (e.g., conservation finance partnerships and extractive industries) but also to serve to translate otherwise weak, unstable and provisional ties between individual actors into far more durable institutional elements of the organisation.[23] It likewise became clear that the instrumental character of the WCC, as a site where the personal associations and ideological work necessary for the renegotiation of organisational order occurs, could be read through a set of analytic devices that build on approaches to analysing the performative aspects of governance: (1) structure, (2) orchestration and (3) spectacle (see Hajer and Versteeg 2005; Hajer 2006).

Structure

Meetings, such as those of the WCC, are built on intent. Organisers envision objectives and desired outcomes, and they seek to manage these in relation to the expectations of their intended participants and a wider audience that will look to the WCC for cues on how to understand the IUCN's position within a broader institutional and organisational environment. The selection of meeting site, the ability and entitlement to attend, the ways in which sessions are organised and designed and the spaces in which they are held[24]—all are a function of decisions made by organisers in relation to their objectives. It is clear that each affects the ways in which knowledge is legitimated and transmitted, and all shape the kinds of interactions that are possible. Paying attention to structure, then, is important to understanding the instrumental qualities of organisations' meetings (e.g., the WCC's).

The organisation of the World Conservation Forum (WCF) is, in some ways, not unlike many academic meetings, although it has more of an 'event-like' quality to it. There are opening and closing ceremonies, for example, but otherwise there are standard plenary sessions and long days divided into sessions of varying types. During the 2008 WCC there were 972 'events' according to Jeff McNeeley, the IUCN's Chief Scientist. Of these, discounting social events, book launches and film screenings, business-related events made up about 10% of the primary sessions. The process for suggesting an 'event' was also not unlike that for an academic meeting. The IUCN Commissions were assigned a certain number of reserved slots for Commission-sponsored events, and the Secretariat structured its own events. Otherwise, members were free to submit proposal abstracts for 'Aliance Workshops',[25] 'Knowledge Cafes', and 'Posters', and they were asked to justify their proposals in relation to one of the Congress's three themes—'A New Climate for Change', 'Healthy Environments–Healthy People', and 'Safeguarding the Diversity of Life'.

A committee within the IUCN Secretariat reviewed proposals, and session organisers were later notified whether their proposals had been accepted. In the case of business-related events, 53% of the successful proposals were put forward by corporations, trade associations or business support units within the IUCN Secretariat. The remainder originated from NGOs and intergovernmental organisations (IGOs).[26] Not surprisingly, many of these proposals were submitted by organisations with existing ties to the private sector such as the Rainforest Alliance Inc., the World Bank, and the World Resources Institute. The substantive presence of business-oriented sessions during the WCC reflects a fundamental ideological shift within the Secretariat, but it also reflects a structural shift in the organisation of the event. In 2008, for the first time in IUCN history, the category 'partner' was included in the list of groups qualified to submit event proposals for a WCC; this led to the inclusion of events that were explicitly organised by private sector actors that were neither IUCN member organisations nor Commission members.[27]

Most of the business-related events fell under the label 'Aliance Workshops'. These, however, were typically organised more along the lines of panel sessions rather than workshops, with presenters sitting at the front of a large room sharing knowledge or experiences with the audience. Other events, given the label 'Knowledge Cafes', were much smaller and typically involved groups of 12 or so people sitting around tables arranged in an open hall. Events were heavily descriptive with little analysis or discussion of the relationship between production processes and the effects on biodiversity, of the biodiversity outcomes of private sector–NGO partnerships, or of the application of 'market-based' mechanisms to conservation.

On the contrary, events were typically performative and celebratory, or designed to promote particular projects, initiatives or 'partnerships'. For example, one session with the not-so-cryptic title of '*Buy, Sell, Trade!*' consisted of participants testing the BETA version of an interactive role-playing game designed to teach people how to engage in biodiversity offset markets. This game had been jointly developed by staff of the IUCN Secretariat and the World Business Council on Sustainable Development

Fig. 4.1 Sessions, such as 'Buy, Sell, Trade!', sponsored by the WBCSD and the IUCN forestalled debate over issues like biodiversity offsets by focussing participants' attention on specific tasks

(WBCSD), under an advisory board of staff from Earthwatch Institute, the World Resources Institute, Katoomba Group–Forest Trends, the US Business Council on Sustainable Development and Fundacion Entorno, as well as several WBCSD member companies (Fig. 4.1).

Another session, entitled 'Transforming Markets: The Private Sector's Role in Securing a Diverse and Sustainable Future', did not deal at all with the constitution of sustainability or diversity; however, it served largely as a green marketing opportunity for the corporate executives invited to present, and as a means to promote the nature of partnerships between those corporations and the session's sponsors: the wildlife trade monitoring network (TRAFFIC), the IUCN and the WWF. The session became a vehicle for the implementation of mutuality and the promotion of partnerships by those already heavily invested in them, rather than a mechanism for evaluating the effect of those partnerships (Fig. 4.2).

Similarly, a 'Knowledge Café', with the provocative title 'The Role of Corporations in Biodiversity Conservation, Corporate Social Responsibility

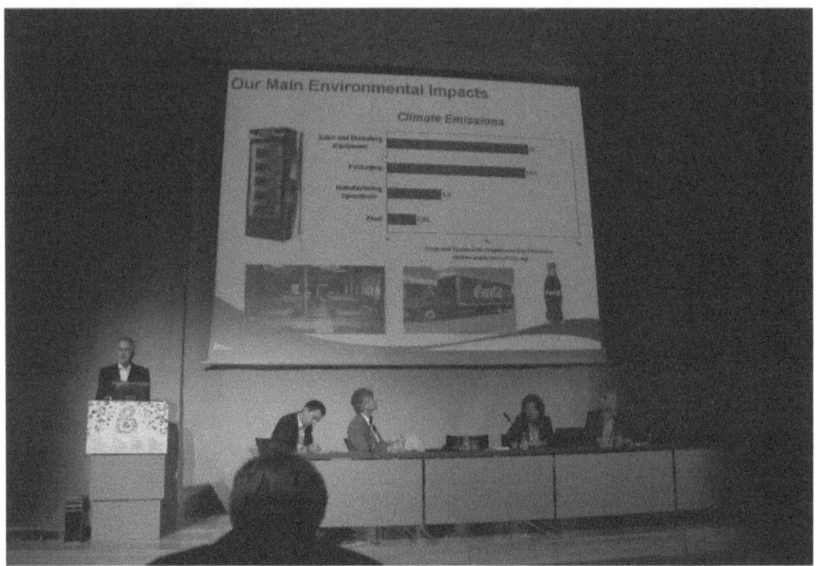

Fig. 4.2 Alliance sessions, such as 'Transforming Markets: The Private Sector's Role in Securing a Diverse and Sustainable Future', served as a vehicle to promote 'private sector engagement'

(CSR) or Greenwashing?', organised by a Spanish environmental consulting firm, became a discussion of the means to access private sector organisations; it barely addressed 'greenwashing' until, toward the end of the session, a graduate student observer intentionally asked a seemingly provocative question: 'Isn't this session supposed to be about "greenwashing"?' (Fig. 4.3). This simple question brought an awkward silence in the group, followed by quiet nervous laughter, but not a response to the question. Again, the audience was a mixture of NGO and private sector employees who were already heavily invested in the promotion of NGO private sector partnerships; they seemed reluctant to question the degree to which NGO–private sector partnerships potentially threatened the capacity of NGOs to claim to be operating in support of the broader 'public interest' in biodiversity conservation.

In addition to regular sessions, there were also ancillary events. The World Business Council for Sustainable Development, for example, hosted two of these. One was the 'Business Night School' organised by the WBCSD

Fig. 4.3 Participants in the 'Knowledge Café'—The Role of Corporations in Biodiversity Conservation, Corporate Social Responsibility (CSR) or Greenwashing?

because 'the private sector thought it was important to share some of our skills and experience in building bridges with NGOs and with governments. We ran 12 sessions ... in business basics, communication, partnerships, and in business impact ..., business-led, business experience-sharing in a very successful program' (WBCSD representative, fieldnotes, October 2008).

There was also a session organised by the 'Future Leaders Team', a program in which young management trainees of global corporations spent four months piloting and testing a tool developed by the World Resources Institute (WRI); it was called the 'Corporate Ecosystem Services Review':

> [A]ll of us took that methodology back into our companies so basically 25 pilot projects within a four-month period which is ..., hopefully, an important contribution to the, to the tool itself, and so that, I guess, commercialization of this ecosystems thinking, this ecosystem services thinking can happen. So we are very happy to be a part of that. The second part of our program involved preparing for the, this World Conservation Forum. And so hopefully you guys have seen the video, the 'Ecosystems: Everybody's Business' video[28] that was prepared by one half of our team. The other half

> worked on a scenarios planning session, which was run on Tuesday of this week. Which really just involved trying to project a vision of the future and get people in the room to think about [what kind of] collaboration was going to be needed between the social side and governments and business. So the tool was scenarios planning and scenarios building, but really the objective was to look at the future and how these three groups within society would have to interact to get things done. (WBCSD Future Leaders Team participant interview, fieldnotes, October 2008)[29]

Aside from the disturbing way in which this young manager has learned to separate 'the social side' from 'business' and 'government', this observation is insightful for the unintended way in which it reveals a commonality that ran through much of the business-oriented sessions of the WCC.

Even though the various types of sessions offered by the organisers were intended to mobilise diverse forms of engagement among participants, they showed a remarkable tendency to bring together actors and perspectives in ways that facilitated the enactment of coherent discursive configurations. The idea of 'balance' achieved by 'working together' referred to by the preceding young manager was particularly notable, and likely not surprising given the 'team-building' exercise he had just been through. Nonetheless, the exclusion of dissent and the absence of any recognition of the ideological basis of struggles over biodiversity conservation that were apparent in the work of the 'Future Leaders Team' was reflected in the remainder of the sessions I attended during the Forum.

During the feedback portion of the 'Buy, Sell, Trade!' session, for example, people gladly provided suggestions on the performance of the game; however, no one raised a substantive question concerning the issues that the game eludes to, such as the efficacy of offsets in conserving biodiversity. These observations point to the ways in which the artefacts of meetings—the use of PowerPoint technology, the arrangement of seats in a room, the format within which interaction is structured, the actions of a Chair—can act to control the circulation and reception of information, encourage the suppression of dissent and encourage the disciplining practise of self-control. But then again the opportunity for engagement, the type of interaction that results, and the possibility of dissent depends, to a large degree, on who is in the room. What is more, at the WCC, structuring who was in the room was a highly orchestrated affair.

One of the most notable aspects of the structure of the World Conservation Forum was the difficulty of navigating through the sessions. Unlike academic meetings, the presenters in each session were not listed by name or affiliation. Sessions were given titles and locations were provided, but it was impossible to know in advance who would be in the sessions and what they would in fact be speaking about. There were no abstracts. The sessions, however, were lent some structural coherence through their organisation into 'Journeys'. They consisted essentially of thematic subprograms—for example, the 'Markets and Business Journey', which assembled all the individual presentations related to business, ecosystem services and the private sector into one coherent program.[30]

Each 'Journey' had 'Guides'; IUCN senior personnel were responsible for 'charting' the journey and managing a staff who made sure things ran smoothly. Two 'Guides' were listed for the 'Markets and Biodiversity Journey'—Joshua Bishop, IUCN's Chief Economist, and Saskia de Koning, listed as Advisor, Business and Biodiversity Programme, but actually a secondee from Shell. The 'Journey' in fact was guided by a host of employees from the IUCN Secretariat, Shell and the WBCSD, along with independent business and biodiversity consultants contracted to help with organisation and reporting.[31] Notably, however, journeys were not simply disaggregated events that pieced together random sessions; they were highly coordinated. Besides, it is this coordination, exposed in part through elements of structure, that provides certain insights into the ways in which events (e.g., the WCC) are orchestrated, the outcomes sought through that orchestration, and how such orchestration can open windows into the ideological and material struggles occurring within the organisation.

Orchestration

Even though a discussion of the WCC's structure might seem mundane, it is in the close attention to structure; the individual events; and the links between them, the organisers, and the participants that we find the markers that begin to outline the contours of association and coordination.[32]

Furthermore, it is in recognising the connections between events that insights into the intent of orchestration become clearer. There is little doubt that the coordinators of 'Journeys' at the WCC had an intent. The 'Markets and Business Journey', for example, was clearly coordinated by people and organisations who had collaborated over the past decade to promote the IUCN's Business and Biodiversity Programme and had served as major anchors in the ideological shift occurring within the organisation. As Frank Vorhies, a consultant brought in to help coordinate the 'Journey' and also the first employee of IUCN's Business and Biodiversity Initiative, put it at the 'Opening of the Business and Markets Journey':

> What we're trying to do with this market's journey, is, we've tried to look at three sectors as priority sectors for IUCN. And then three issues related to the economy. The three sectors are agriculture, tourism, and extractive industries. ... Plus we're looking at three issues. One of the issues is partnerships. The second we call incentives. You might call it measures, instruments, regulations. And the third one is indicators. The first one on partnerships, we're looking at ways in which the IUCN community can work with business and with business actors. The second one on incentives is the IUCN community working on the regulatory and the voluntary frameworks in which businesses operate. For example, international payments for ecosystem services is being talked about today. The third one on indicators is an interesting one for the business journey because how can we take the information in the traditional IUCN community, protected areas indicators, ecosystem resilience indicators, indicators on the status of species, species survival and so on, and use those indicators as instruments for, for business to be better in their biodiversity performance. So you see, three sectors, three themes of economic indicators, and hopefully if you go through some of these events you will come out with a better understanding of how IUCN's community is and could be addressing conservation through addressing market processes, regulation of market processes, business practises and business standards, and so on.

What Vohries does not say in this is that the 'Markets and Business Journey' is also a coordinated exercise, which had been constructed through existing associations between actors within the IUCN Secretariat, private sector organisations and government agencies, to extend models of private sector engagement and, as such, was specifically designed to draw actors together in an effort to generate associations around the innovations he describes. A look at the schedule for the 'Journey' clearly delineates the kind of channelling intended by the organisers, as does the description from the journey programme[33]:

> If you seek a better understanding of the challenges and opportunities of enlisting business and consumers in biodiversity conservation, join us on a journey through the markets!
>
> Throughout the Congress there are many events related to economics, markets and business. Our journey takes you to a cross-cutting selection of these, focusing broadly on major market sectors and key market tools.
>
> Our journey highlights several events related to agriculture, extractive industries and tourism—all of which have significant linkages to biodiversity. Our journey also highlights events related to partnerships between conservation

> organizations and business, new economic incentives and biodiversity indicators for business. These events all profile innovative approaches to using the market to conserve biodiversity.
>
> When you need a rest, you can always stop off at the World Business Council for Sustainable Development Pavilion and learn more about what the WBCSD member companies are doing to improve their environmental performance. You can also stop off at the Futures Pavilion where fellow travellers will also be congregating to discuss market challenges and opportunities for IUCN.

In some ways, the concept of 'Journeys' is akin to 'threading' in online computer conferences or forums. They work through sorting 'posts' into coherent threads to eliminate confusion from competing topics (Kean 2001). Nonetheless, 'threading' relies on the existence of an online archive through which users can, after the fact, read within and across threads to gain a fuller sense of the 'discussion'. The same is not true of channelling tactics (e.g., 'Journeys'). Participants are corporeal and limited to the cognitive space of their body. They either attend a session or they do not. If they miss a session, there is no archive to which to refer. To the extent that 'Journeys' are designed to 'channel', they orient like-minded people along a path of least resistance that allows them to engage with others in ways that minimise conflict and dissent. It was obvious in several ways that the IUCN leadership anticipated the possible effect of this channelling.

The Director General's address, for example, just before closing the opening plenary, described the conference 'Journeys' but encouraged people to engage in what she called 'creative collisions' and to get out of their 'comfort zone during these days of the Forum'. She continued:

> It's really important to move in between the different streams and the discussions that arise in them. We depend, IUCN and all of you depend on the cross-fertilization of ideas and experiences and solutions that this forum is going to offer to you. So, this is the only way that we will save the world, change the world, which is really why we're all here with our optimism and energy. (Fieldnotes, Barcelona, October 2008)

Nonetheless, those 'creative collisions' were hindered in a number of ways. 'Journeys' at the WCC had coordinators, they had booths where participants could congregate, and they had opening sessions in which participants were encouraged to follow the 'Journey' schedule and to provide feedback to the coordinators. Interaction between fellow 'Journey' members was encouraged.

Identifying markers were handed out, participants were encouraged to follow specific 'Journeys' by brochures and events programmes that identified the time and location of each relevant event (as opposed to the full Congress Programme that was only available online), and they were encouraged to socialise in 'rest areas' explicitly connected to each 'Journey' (Fig. 4.4). All of this created a tendency for each of the sessions to encourage participants to think of themselves as part of a coherent group. During the 'Opening of the Markets Journey', for example, Saskia de Koning gave a markedly different message than the Director General:

> [In the Markets and Business Journey programme] You'll find the highlights of the program and the events that we have selected for you; various events around markets and business. But, of course, there are a lot more events related to this topic. But if you want to see those you have to use your guide and find them. The other thing that we have done is we have made this button—you can see them—'Nature Is Our Business', but what we want, is we want to create a community related to markets and business so if you might take a pin and have a pin—and you're more than welcome to grab one—and please leave your business card if you can, it would be very helpful to us. So that you will find that if you wear this pin you'll find other people who are part of this journey and you'll be able to recognize them. If you need a rest, like you are doing right now, if you want to stop, if you want to have a break, come over here and you'll find people who are interested in markets and in business, or over there [pointing to another seating area across the Hall] you have the pavilion of the World Business Council on Sustainable Development. You've probably seen that already. They also have people who have an interest in markets and business.

The message here is clear: '[W]e have prepared a path for you to follow; we have prepared markers that signify an identity, so that you can identify, and identify with like-minded people; we have provided space for you to socialize with like-minded people; and we want to translate personal ties into durable associations that can be sustained after the Congress'. Indeed that seemed to be the effect of the 'Journey'.

Following and observing the 'Markets and Business Journey', I consistently found myself in rooms with many of the same people because they followed the journey programme from session to session. In the sessions I attended, there were little if any contentious comments.[34] This was facilitated, intentionally or not, by a programme that failed to identify speakers or subjects of presentations. Not knowing that a senior executive from,

Fig. 4.4 The cover page of the 'Markets and Business Journey' suggested a reworking of the old English idiom 'Money does not grow on trees'

for example, Coca-Cola or Shell was going to be giving a presentation in a place at a particular time, simply made it much more difficult to target protest, or to challenge claims that they might make during their presentations. In addition, this absence of dissent made it much easier for participants in the 'Journey' to conceive of themselves as part of a 'community'.

If we define orchestration as the coordination and the intentional act of organising individual elements of larger assemblages such that those individual elements work toward a predefined endpoint, then it seems clear that the organisational tactic of arranging sessions into 'Journeys' was an element in a clear strategy of orchestration—a strategy that involved coordination over time and space outside of the physical bounds of the meeting space, the channelling of interaction and discussion, and the scripting of narratives and configuration of space in ways that enhanced the likelihood that the desired endpoint would be reached. But in a meeting like the WCC, orchestration occurs across scale, and it is productive to ask how the 'Markets and Business Journey' and its coordinating themes, link directly to the broader renegotiation of the organisational order of the IUCN that I alluded to earlier.

In some ways, this renegotiation of organisational order is already clear. When conservation is treated as organised political practise, the empirical evidence showing the degree of 'private sector engagement' with the IUCN Secretariat, the conflict among IUCN members around that engagement, and the actions of both the Secretariat and the Director General in pursuing new paths of private sector engagement firmly set the tone of the Congress before the meeting even convened. From the opening ceremony onward, one after another, in major addresses during the Congress, the IUCN's senior leadership, in their tone as much as in their words, sent out the key message of embracing business that James Kantner very quickly picked up on in his blog in which others participated:

> Julia Marton Lefevre (Director General): If we want to influence, encourage and assist the world to change, we can only do so when we work together in an efficient and seamless manner, and allow **new partners** to join us.[35]
>
> Bill Jackson (Deputy Director General): With **biodiversity as our business** we have new people, ideas, but we also have resources.[36]
>
> Jeff McNeely (Chief Scientist): We've also demonstrated the value of reaching out to **new partners**. Having 7000 people here from many different parts of life has been a very powerful indication of how popular our issues are becoming. Following calls from you, our members, to **work with the private sector**, the forum has attracted the expertise, the resources, the perspectives from numerous parts of the private sector. …

> Joshua Bishop (Chief Economist): **We need to bring in the private sector**, the consumers, the business community generally to support conservation and ecosystem management. And we can't do it alone[;] we have to work together to make this happen.

In these remarks, the senior leadership of the IUCN Secretariat sent a coordinated and clear message to the membership and the private sector regarding both the role of the private sector at the Congress and in the future direction of the IUCN.[37] To legitimate their engagement with the private sector, it is important to the organisation that a clear message on the direction they would like to pursue be communicated to the membership. But to solicit those new partners, it is also important to the Secretariat to convey a sense of a coherent membership that uniformly agrees on the direction. This message was also clear in the addresses of the IUCN executive, but it was articulated in a more hesitant tone. For Jeff McNeely, the WCC was an:

> ... illustration of the merging of members, commissions, the Secretariat and partners. Truly a celebration of the one program approach that we've all been advocating ... different events sort of merged together, they float together in a seamless action of ideas demonstrating yet again the essential unity of what we all stand for; the importance of not drawing boundaries around the work that we do but **bringing it all together**. Conservation has been demonstrated again as a unifying force, not a collection of silos standing in lonely splendor.

For Bill Jackson, 'one of the outcomes of nearly every session of this conference was a sense of how the urgency of our situation was no longer seen as isolated by geography or gender or class or education. **We are all in this together'**. Julia Marton-Lefevre noted:

> [A]s environmentalists, our time has come. The level of awareness is higher than it has ever been. And while some just talk green, nobody can ignore environmental concerns now. So, in a way, yes, our time has come, but we also have to seize this occasion. We can always do better, **but let's recognize together that we are moving in the right direction**.

Orchestration, then, extended far beyond the sessions and the 'Journeys', which were to some extent reflective of the internal politics of the IUCN Secretariat that saw different units jockey for position and authority. It also was reflected in the intentional, prescripted and prescriptive narratives of the IUCN's senior leaders who sought to communicate the direction of the organisation, to set the tone of the meeting, and to establish the basis for a 'new normal' within organised conservation. The message from senior leadership regarding engagement with the private sector was clear in their remarks. Yet, for orchestration to gain influence, it requires not just an audience but also a stage. It requires the production mechanisms of the meeting: the opening ceremonies, plenary addresses, and the welcoming of dignitaries and celebrities. This is the realm of spectacle.

Spectacle

Research on the analytic utility of spectacle and celebrity for understanding the cultural politics of contemporary conservation started in 2010 (Igoe et al. 2010); it is useful to think of theoretical approaches to spectacle in relation to the work being accomplished by the World Conservation Congress simply because so much of the WCC is grounded in spectacle and performance. Debord (1967), in his original work on the role of spectacle in mediating social relations, made several proposals including the following: spectacle imposes a sense of unity onto situations of fragmentation and isolation, and spectacle is an omnipresent justification of the conditions and aims of the existing system (i.e., it is an instrument in processes of ideological and material domination that conditions people to be passive observers).

From Debord's perspective, the WCC was nothing, if not spectacular. The configuration of visual and aural experiences clearly were designed to impose a sense of unity on a fragmented audience. The foyer of the conference centre was filled with a series of pavilions, each with their own speaker's platform, sitting areas and video screens. Images of nature abounded, but so did a prevalent visual message: As participants entered the convention centre they looked straight ahead at an image of the globe littered with the names of what conventionally are thought of as some of the most environmentally damaging companies in the world—Shell, BP, Dupont, Newmont Gold, Dow Chemical, Chevron, Mitsubishi. This was the pavilion of the World Business Council on Sustainable Development. To participants' right was the visibly trademarked 'Red List™' printed across the wall of the IUCN species pavilion. To their left was IUCN's Forest Conservation Programme pavilion with a prominent banner in the foreground proclaiming: 'Private–public partnerships can achieve sustainable and equitable development'. The mediating signs of a renegotiated organisational order set the tone of the Congress even before participants set foot in meetings or in the plenary hall.

It was in the plenary hall, however, that the mediating effect of spectacle took over. The first evening of the Congress began with an opening ceremony steeped in references to a feudal past as the presence of European Royalty was announced to an assembled audience who were then made to wait while they were 'received' outside of the main hall. The seating of Royalty was followed by a procession of VIPs, set apart by the red neck straps attached to their identification badges (to set them apart from the green of regular attendees). The ceremony itself made a nod to Royal courts of the past as musicians accompanied 'larger- than-life' video presentations on the main stage, and 'players' planted in the audience 'magically' appeared under soft spotlighting to 'debate' various ideological approaches to what one of them called 'the environmental crisis that we all face'—their voices reeking with earnestness, suspicion, contempt and crisis.

Then, tumblers whorled across the stage and acrobats spun gracefully as they climbed and wound themselves around five long red cloth banners hanging from the ceiling. One of these was turned after each performance to reveal a reverse side emblazoned with a word, gradually exposing the sequence: SUSTAINABILITY, AWARENESS, EQUITY, BIODIVERSITY AND ACTION. Celebrities, such as Mohammed Yunus and David Attenborough (by video), and 'lesser' figures (e.g., the Spanish Minister of Environment) appeared on stage to address the audience, assuring them that they were doing essential work.[38] In addition to all of this, the orchestrated narrative of the organisers lay under the surface: 'we' are all in this together, 'we' must 'move forward', 'we' must work with the private sector.

The next morning's opening plenary was equally spectacular. Ted Turner served as the key celebrity[39]; an IUCN partnership with Nokia was highlighted as a way to use mobile communication technology to engage youth in the environmental movement[40]; and a *faux*-debate between panelists, moderated by a South African television host and former CNN International anchor, Tumi Makgabo, was clearly designed to articulate the key conservation conflicts as they were understood by the Secretariat leadership. Indeed, this 'debate' is a useful empirical device in an analysis of the work accomplished by spectacle. For as much as spectacle serves to mediate, the combined use of sound, image and space that constitute spectacle also must be mediated and orchestrated.

In this case, the use of a television host as 'moderator' of the 'debate' was wholly appropriate because the entire event felt like a skilled television production, and watching the 'debate' unfold on stage was much like watching a television program. Indeed, the entire event came off like a talk show and likely was intended to. In addition, similar to a television program, in which the medium itself defines the limits of interaction, it was made clear from the beginning that the audience at the WCC was there simply to view. The panelists, prompted by the scripted questions of the moderator, could say what needed to be said, and any thoughts or input that the thousands in the assembled audience might offer were of little value. There would be no substantive debate, the audience members were not allowed to engage with the panelists, and the moderator controlled the flow and posed the questions. She made this clear from the beginning: '[U]nfortunately, there is no time to open the floor for dialogue but the experience on stage is enough to answer any question'.

A structured format had been developed and deviation from that 'script' implied excessive risk to the production quality of the spectacle. It also threatened the ordered messaging of a narrated script that ended on a very specific note put forward by Achim Steiner, former Director General of IUCN and then current Executive Director of UNEP, and Bjorn Stigson, President of the WBCSD: '[B]usiness is part of the problem and part of the solution'. This is a standard line in the rhetoric of business and biodiversity engagement that quite obviously obscures diverse ideological and material interests between conservation interests and business. But far more important is where this line sits in terms of orchestration. This was the last message that emerged from the opening plenary, leaving people with the role of business in biodiversity conservation on their minds as they went off to engage with the rest of the forum.

Here is an example of the alignment of structure, orchestration and spectacle. The panel was positioned to be one of the key agenda-setting items of the Congress because it drew most of the participants together in a common meeting place. Besides, although it had the appearance of a debate, it was effectively scripted in a way that privileged the role of business in biodiversity conservation as one of the key themes of the Congress. The effect of this scripting was reinforced by production practises—the mimicking of a television program—that imposed a sense of unity on an isolated and fragmented situation by isolating interaction among actors on the stage from actors in the audience. This emphasised the performative aspect of the 'debate', bringing out the latent qualities in an audience accustomed to adopting a docile position when interacting with the medium of television or theatre, and assuring greater control over the messages emanating from the stage because the audience was precluded both by instruction and proclivity from introducing dissent to the scripted narrative.

Nevertheless, what is particularly striking is that all the large plenary sessions of the Congress were orchestrated in similar ways in which moderators, often the Director General, acted as brokers for the audience—that is, transitioning them away from the 'show' of the discussion with the help of a musical and visual interlude, a commercial and then bringing them back in with introductions to subsequent events. Through all of this, the audience sat politely, applauded at appropriate moments and exercised the self-control and self-discipline characteristic of meeting culture. We, as an audience, were part of the spectacle—spectators watching a program, consuming celebrity in the guise of 'expertise'. But more significantly, we were—in an environment that encourages passivity and reinforces the exercise of self-control and self-regulation—consuming the symbols and narratives designed by the conveners to communicate and justify the new aims of the organisation and changes in the organisational order.

Structure, orchestration and spectacle, as analytic devices, are obviously intertwined. In a world in which image and volume are hypervalued, spectacular domination acquires a capacity to convey authority and aids in securing legitimacy and position within institutional power relations. Meetings, such as those of the WCC, have become, in many ways, a theatrical stage on which actors play out roles designed to advocate positions that they believe are important in the repositioning of the organisation and its ability to secure greater resources. In this world, the legitimation of actions and authority increasingly requires spectacle. In addition, organisations like the IUCN strategically use spectacle to attract the attention of media, circulate self-representations and, hopefully, their message, as well as to secure the interest of the private sector.[41] They do so in the belief that engaging certain private actors not only will confer legitimacy with superordinate institutions (e.g., the GEF and UN agencies) but also extend the media presence of the organisation and its capacity to secure new 'partners'.

This is one of the reasons for inviting keynote presenters (e.g., Ted Turner and Mohammed Yunus) to speak in an opening ceremony.[42] They can tell the assembled audience nothing about biodiversity conservation, but their presence can communicate 'importance' and provide access to new sources of material support and the credibility desired by IUCN's leaders for both personal and organisational reasons.[43] Ironically, however, the production of spectacle is an expensive proposition and the IUCN consistently has trouble securing private sector funds to actually mount the WCC. As one consultant put it to this author, a senior IUCN staffer came to him in the months leading up to the Congress and confided 'you know we've got these two pavilions, climate change and energy and we don't have sponsors for them. We're in trouble for the Congress. We wanted to have these pavilions, and we don't have any sponsors. Can you help? If we brought you in on a contract could you help us with these pavilions?' (Fieldnotes, Bonn, Germany, May 2008).

This relationship between spectacle—designed to secure the interest and the material support of the private sector and to communicate the message of shifting alliances within the organisation—and orchestration needs to be made clear. The orchestrated narrative, one in which IUCN's senior leaders work to convey the impression of a coherent organisation with uniform objectives on which all members of that organisation agree, is meant to disguise the degree of dissent that exists within the IUCN—the ideological and material struggles that take place on an ongoing basis.[44] The point here, however, is that the desire on the part of the Secretariat to convey the impression of a union with common purpose, and uniform agreement on the means to achieve that purpose, is designed to appeal to a corporate model of business management that brooks little internal dissent, especially if it is publicly expressed.

In many ways, that model does not understand the kind of debate and ideological and material struggle that legitimately goes on within the IUCN, which is, after all, a Union. Rather than read it as a sign of healthy participatory engagement in a debate over organisational direction, they see it as a symptom of dysfunctional leadership. As one oil executive casually said to me when I asked him about his impression of the meeting: 'IUCN's a dysfunctional organization, they don't have their act together'. When I asked why he thought this was so, he replied: 'If they had their act together they wouldn't allow a motion to terminate their agreement with Shell'.[45]

It is this desire to secure and engage 'new partners', and to fend off the 'risk' evident in the executive's remarks, that drives the coordination of the script asserting commonality with an IUCN membership, and that fuels the (highly energy-consuming) development of spectacular conservation. Those IUCN leaders engaged in performance know that they are acting; they know that in asserting a commonality, they are engaging in deception.

They also know, however, that frank and open debate and the ongoing presence of dissent within the organisation does not secure legitimacy with those 'new partners' they seek to cultivate. From a corporate perspective, ongoing dissent within the organisation is considered a risk—a potential source of liability and instability—and an ongoing threat to the pursuit of organisational objectives. Besides, in a severely restricted funding environment in which private funds have become increasingly important to conservation organisations, it also has become a significant liability for the IUCN (cf., Brechin and Swanson 2008).

Orchestration and spectacle, then, both serve to reinforce the self-regulation and self-discipline of conditioned behaviour typical of meeting culture and of effectively limiting dissent through the application of structure. It was no accident that the plenary sessions that brought thousands of IUCN members with vast accumulated experience in biodiversity conservation together in the World Conservation Forum provided no opportunity for those members to speak. It was not accidental that the names and affiliations of speakers and the times of their presentations were not listed in the Congress programmes.

In addition, it was not accidental that the opportunities for those members to speak were restricted to random encounters in hallways or in meeting rooms during sessions that were organised into 'Journeys' designed to channel and contain Congress participants. No doubt some individuals broke out of these 'Journeys' and some sessions did engage in substantive debate, but this involved work and added energy as if 'swimming against the stream'. All these elements of structure and orchestration required conscious deliberation, decision and intent by actors in the IUCN Secretariat and their broader network of associations.

One way, then, of interpreting the WCC is as an expression of the intent of the IUCN Secretariat, but that would be a mistake. It is more accurate to look at it as an outcome of struggles within and between organisational actors, including various units and actors within the Secretariat, Commissions, the state, and NGO members and actors (e.g., the private sector) that see the IUCN as an organisation sufficiently important, either as a threat or a resource, that they feel the need to attend to it. That is the value of focussing to the WCC as a site and object of study; for as much as it can be treated as an important vehicle in organisational change, it also serves as an opportunity to witness the development of associations and to weigh what was said and acted on by individual actors against later representations of the event.

Given that the WCC can be read as a reflection of struggle within the organisation, and that there was substantive evidence at the WCC of membership opposition to the way that 'private sector engagement' was being pursued by the Secretariat, the way in which structure, orchestration and spectacle combined to legitimate 'private sector engagement' suggests that substantial ideological shifts are occurring within the organisation. In addition, attending to these changes could provide a greater understanding of how the organised political project of conservation is likely to unfold in the near future. Much of this was in fact foreshadowed by sections of the WCC in which it became apparent that spectacular domination, the same as hegemony, can never be complete.

The Incompleteness of Spectacle and Orchestration

Notably, the WCC does contain a component through which it is difficult, though not impossible, to invoke spectacle, employ orchestration and manipulate structure. This is the Members' Assembly, which is structured by statute, unlike the Forum, and where discontent, dissent and the significant degree of tension between IUCN members and the Secretariat become readily apparent. This does not mean that the Secretariat, in association with certain members and partners, did not try to use the Members' Assembly to maintain their push for a new organisational order. In her opening address, for example, the Director General continued in the tone she established during the Forum:

> We can take the known path, a familiar and comforting one, where we discuss and argue among ourselves. Where we do interesting work, raise red flags and have a moderate influence. Or we can take the other road, the proverbial 'less-travelled' one, where we embrace rather than resist change. It is a road where we travel with **new partners** because they represent our reality as parts of the society we were set up to **influence**, **encourage** and **assist**. It is a road where IUCN builds bridges – not necessarily consensus – between diverging interests. It is one where IUCN leads a massive movement, which is the only way to affect the kind of changes needed to save the only planet we have. We don't need to change our mission statement in order to achieve this. We just need to realize that progress is not always linear; that we have to be smart and strategic in the way we make the case for Nature to an ever-widening audience; and that we are so much stronger when **we work together**.

Nonetheless, the Members' Assembly is also where members can speak, within the rules, and challenge this renegotiation. During the 2008 Members' Assembly the overarching point of contention surrounded a number of motions related to the way in which the IUCN Secretariat had engaged with the private sector since the 2004 WCC; primary among these was a motion to terminate an agreement that the IUCN had struck with Shell.[46] Even though the behavioural qualities of 'meeting culture' held during the Assembly—with members waiting their turn to speak and demonstrating at least minimal respect for the views of others—there was a definite loosening of self-regulation and self-discipline, particularly in the contact groups designed to establish whether consensus could be reached on the text of motions.[47] Some supporters of the resolution made a clear point that although the Secretariat was worried about its reputation with business if the motion should pass, it did not seem worried about its reputation and obligations to the membership of the Union.

Ultimately, the motion was defeated, even though it secured the support of more than 60% of the membership. The reason for the failure is revealing. The IUCN operates on bicameral principles. One 'house' of the membership is made up of NGOs, while the other 'house' is composed of state members. According to statute, for a motion to pass, it must secure a simple majority in both 'houses'. Despite support from some states, the 'Shell motion', as it came to be known, was voted down by a clear majority of state members. In many ways, the failure of this motion encapsulates the way in which the WCC exposes to view insights into the renegotiation of organisational order within the IUCN that would otherwise be hidden. Much of this change is contained in the following simple unwritten statement made by Joshua Bishop, the IUCN's Chief Economist, at the close of the 'Markets and Business' Journey, hosted at the pavilion of the WBCSD:

> We call this the markets and business *journey* to remind people that four years ago in Bangkok we had a markets and business *theme*, and even before that the business community was involved in IUCN Congresses and General Assemblies. And I think I can safely say that it's probably now a permanent fixture; that markets and business are not going to go away in the IUCN agenda and are likely to go from strength to strength. (Fieldnotes, Barcelona, October 2008)

One effect of the shift in organisational order, which, ironically, also contributes to facilitating even greater change, can be found in the reactions of some members, particularly Friends of the Earth International (FOEI), a primary sponsor of the Shell motion. The FOEI withdrew its membership in IUCN shortly after the close of the conference, citing their 'belief and experience that partnerships between transnational corporations such as Shell and conservation organizations such as IUCN have a disadvantageous effect on community struggles to protect their environment …' and their disillusionment with a process that allows states to block the vote of a majority of the IUCN membership (Bassey 2009). The remarks of the IUCN's Chief Economist and the departure of the FOEI are indicators that the WCC is an effective vehicle in the renegotiation of organisational order, allowing for declaratory statements and performances of intent and achievement that establish ideological boundaries of 'belonging' for the membership of the organisation. The FOEI withdrawal, however, is also a sign that the internal contradictions of organisations (e.g., the IUCN) become more apparent and visible through pressures exerted by their external environment.[48] Among other things, this departure serves as a sign that 'the restraints, compromises, and accommodations defining the institutional core and rationale' of an organisation are subject to failure in networked organisational forms that can dispense with much of the highly formalised systems at the core of bureaucratic organisations (Reed 2001: 142).

Concluding Remarks

Conservation typically is treated as ontological, both as practise and as an ostensible movement. At sites, such as the World Conservation Congress, leaders of conservation organisations insist that this movement is constituted by the force of some set of relationships oriented toward a common objective, and that it is into this movement and its common aims that others, such as the private sector, must be brought. What becomes clear through turning the ethnographic lens on conservation organisations and institutions, however, is that conservation, in practise, is defined through association—that is, the association of individuals, organisations, institutions, bodies of knowledge and interests—that often becomes most clear in the wake of new innovations. In many ways, 'conservation-in-the-making' is visible only by the traces left as these new associations are being shaped—as the implications of innovations for various actors become clear, and as personal ties are institutionally reconfigured into more durable associations.

Events constitute the political sites where much of that reconfiguration is rendered legible, such as at the WCCs; where the political future of conservation is negotiated; and where struggles over deciding what binds 'us' all together are acted out. It is where, for example, the tension over the articulation of market-based mechanisms of conservation practise becomes visible; where it becomes possible to watch the personal social ties that have been formed around 'private sector engagement' being converted into much more durable institutional arrangements. In other words, to become, in Bishop's words, 'a permanent fixture', no longer so heavily reliant on the vagaries of personal relationships, but written into the standard operational practise of the organisation.

This chapter has suggested that this performative work is predicated, in part, on the act of meeting and the ways in which meetings serve both as sites for the formation of associations and as vehicles that can be structured, orchestrated, and represented in ways that privilege certain positions in renegotiating organisational order and accommodating the presence of new actors. In many ways, then, the WCC can be treated as a site in the neoliberal restructuring of conservation. Meetings are sites at which organisational agency can be seen to work toward achieving and articulating the configuration of new organisational order, and at which the interests of capital accumulation receive an unparalleled degree of access and consideration in conservation planning and practise. In an effort to secure institutional legitimacy, events have succumbed so much to the logic and mechanics of spectacular domination that they require the material resources of the private sector simply to come into being.

Although these empirical insights are helpful in contextualising the organisational and, therefore, the operational direction of biodiversity conservation, an equally salient outcome of this work is the way in which it reveals ethnographic practise as a useful means to comprehend events (e.g., the WCC). This is in addition to the normative behaviour they invoke, as a central instrument in the configuration of the organisational order of a movement like conservation in relation to dominant political projects, like capitalism.[49] Because large meetings have become a key mechanism in the negotiation of transnational environmental governance, and as the stakes in the form of that governance have increased, meetings have become sites where struggles over the configuration of biodiversity conservation practise play out, *and* a vehicle that can be used to favor certain outcomes in those struggles.

The work this chapter discusses also situates the WCC as one node in a broader field site, the spatial and temporal dimensions of which are defined by tracking the relationships that serve as the basis for associations. That such meetings are increasingly serving as sites of struggle over conservation policy and practise, and have come into being as a function of emerging associations, positions them as important sites for ethnographic fieldwork. It also highlights a need not to focus on congresses and/or conventions as isolated events but as devices in a repertoire of mechanisms involved in the formation of associations that subsequently acquire the strength and durability to alter organisational order. After all, ideological shifts within organisations occur incrementally. Although congresses, conventions, conferences and similar events serve as crucial sites to witness the interests behind such incremental shifts, the associations that facilitate them, and the ways in which meetings can be used as instruments to achieve incremental shifts, while creating the impression that these shifts are consensus-based, the only way to acquire a comprehensive view of this process is to situate events in relation to broader political projects of capitalism and the state.

The WCC, nevertheless, was not a 'neutral' event constituted in a vacuum of agency. It has purpose and intent, assigned by individuals differentially situated within, between and beyond the IUCN Secretariat, its 'partners' and member organisations. In addition, there is little doubt that the meetings can be interpreted as a reflection of the current configuration of ongoing ideological and material struggles within the IUCN and, by extension, the conservation 'movement'. According to one IUCN senior staff member interviewed during the Congress, the shift to market-based dialogue is dominant in the IUCN's decision-making bodies; there is substantial support for 'private sector engagement' and market approaches; and a group's visibility and credibility within the IUCN can be enhanced by aligning with those decision makers, particularly the Chief Economist and staff dealing with markets and businesses.

This is an important insight because it points to the ways in which the WCC can be read as an expression of influence and authority not just within the IUCN but also between it and more dominant actors in its institutional environment (e.g., GEF and UNEP). Recognising these translocal relationships and the ways in which associations form in those translocal contexts helps to understand their relative influence in biodiversity conservation policymaking. The 2008 World Conservation Congress provided a notable window into the consolidation of such relationships, perspectives and processes and their role in shaping the new organisational order of the IUCN; one that situates markets, business and private sector actors firmly at the core of the Secretariat, if not the IUCN membership.

This chapter's author does not want to overstate the importance of meetings as field sites. There are many locales in which the reconfiguration of organisational order might be studied. Long-term fieldwork within a conservation organisation, for example, would undoubtedly yield useful insights into the interests behind incremental shifts in organisational priorities. The point here is not to privilege any one field site over another, but to highlight the ways in which the ethnographic study of meetings across time and space offers the opportunity to longitudinally track and document the actors, interests and processes involved in the renegotiation of organisational order; to identify the ways in which associations that give rise to new innovations stretch across a broad range of conservation actors and organisations; and, ultimately, to improve the ability to link those processes, across scale, with ecological dynamics at particular sites.[50]

As innovations proliferate in conservation, the boundaries of the 'movement' shift and, for some, create uncertainty about their position within that 'movement'—the 'conservation community'. Questions of belonging arise and assertions of crisis are lobbed. Those who once thought of themselves as being at the core of effective action, find they are marginalised, and those who once thought of themselves as excluded become central. In many ways, the introduction of new innovations to conservation organisations, and the crises of identity they invoke, exposes the degree to which those organisations are sites of ideological and material struggle, and the idea of 'conservation' fraught with conflict. To explain and understand the implications of this struggle, it is important to be present as the relationships and associations that have constituted active conservation over the past decades shift and reassemble (as they always do); to engage in what Latour (2007: 11) calls 'the costly longhand of associations'; to track the processes and actors involved in the reassembly of conservation; and to pay attention to events and spaces, such as the WCC, where one can track the shape, size, heterogeneity and combination of associations engaged in producing new fields of conservation.

Notes

1. The World Conservation Congress (WCC) is divided into two primary sections: the World Conservation Forum (WCF) and the Members' Assembly. The WCF is somewhat akin to an academic conference with a variety of organised panel sessions and roundtables and is accompanied by a trade show of sorts in which conservation organisations and related enterprises display their 'wares'. Following the closure of the Forum, which is open to non-IUCN members, the IUCN Members' Assembly convenes to elect new officers, receive reports from the Secretariat and vote on member-sponsored motions designed to direct IUCN policy and practise. The motions' process also involves the formation of contact groups around controversial motions in an attempt to reach consensus on a final text before the motion is put to a vote. Historically, access to the Members' Assembly has been limited to IUCN members and Secretariat staff. But at the 2008 WCC it was opened to nonvoting observers.
2. James Kantner is a primary contributor to Green Inc., a blog on NYTimes.com, the website of the *New York Times*, and was the sole reporter for the *Times* covering the WCC. Of the eight posts he submitted from the meeting, five related to the role of business in addressing environmental issues. Two of those were interviews with the CEOs of Rio Tinto and Shell. There were no interviews with conservation or social scientists, and no coverage in the Science section of the newspaper.
3. Of course, there is always the possibility that the narrative was prescripted and developed in conjunction with certain WCC participants. Notably this emphasis reflects Kantner's background as the *International Herald Tribune*'s primary correspondent for *European Business Affairs* and what he calls the 'business of green'.
4. By 'modernist conservation' I mean the policies, programs and projects of large international conservation agencies and national governments. This is not to assign any priority to this work but to distinguish it from the many small-scale conservationist practises that fall outside of this domain.
5. Notably the willingness to expose these 'engagements' is geographically variable. For example, an extremely well-known international environmental organisation, which is open to advertising their relationship with a dominant U.S. soft drink manufacturer, is very reluctant to openly promote it in Europe because of a feared backlash from what it perceives is a more 'radical' and anticorporate membership base (Fieldnotes, Barcelona, Octobrt 2008).
6. Since renamed the International Council on Mining and Metals.
7. Steiner has since become Executive Director of the United Nations Environment Program (UNEP).

8. Dissent was much more apparent during the Members Assembly as some NGO members accused the Secretariat of having ignored the constraints the membership placed on IUCN engagement with the private sector through resolutions passed at the 2004 WCC. Several members I spoke with at the 2008 WCC claimed that this was evidence that the Secretariat did not feel bound by resolutions passed during the Members' Assembly, felt free to pick and choose the resolutions they would act on and those they would ignore, and would continue to expand private sector engagement regardless of resolutions adopted by the membership.
9. This has become common practise at the IUCN, which now has secondment agreements with Holcim Cement, Shell and the WBCSD. Indeed, one of the organisers of the 'Markets and Business Journey' at the WCC, who was listed as 'Advisor, IUCN Business and Biodiversity Programme' is actually a Shell employee seconded to IUCN for two years.
10. This includes groups, such as the Sierra Club, that, while critical of IUCNs agreements with oil companies such as Shell, have established their own partnerships with chemical product manufacturers (e.g., Clorox). Notably this agreement became an element in a discursive struggle between camps at the WCC with supporters of the Secretariat's position on private sector engagement instrumentally referring to the Clorox partnership to question the credibility of Sierra Club-sponsored criticisms of IUCN partnerships.
11. Part of the reason for this failure to grapple with the intersubjectivity of institutions stems from the theoretical and methodological 'path dependency' of the early work in regime analysis (see Young 1982, 1989, 1994). It is notable that, even as organisational and institutional ethnography was becoming prevalent in anthropology and sociology, regime analysis was slow to recognise the theoretical and methodological benefits of an ethnographic approach to understanding environmental governance. Some of the reasons for this can be found in disciplinary policing that accompanies 'path dependency'. In her own interview-based work on knowledge production in the International Monetary Fund, for example, Martha Finnemore felt that some of her colleagues saw her depth-interview approach, let alone ethnographic fieldwork, as methodologically radical (Finnemore pers. comm. 2004; see Barnett and Finnemore 2004).
12. Notably in Finnemore's analysis she continues to adhere to textual analysis and to avoid engaging in sites of interaction in which discourse is not only produced and deployed, but in which the production of intersubjective meaning might be observed. Her focus on intersubjective understanding is also problematic as it avoids engaging with the very real social controversies that are involved in the process of negotiating institutional and organisational order, and in analysing how meetings are structured to facilitate intersubjective understanding as a mode of masking those controversies (*cf.*, Nader 1995)

13. Vogler's work represents the very late coming of regime analysts to social constructivist perspectives on nature, again pointing to the constraining quality of disciplinary perspectives.
14. For example, in work that I am undertaking on the Convention on Biological Diversity (CBD), being present at a series of meetings and getting to know activists and state delegates makes it easier to identify when NGO delegates are sitting at the delegates table and acting as advisors to state delegations and to track these associations through the process of negotiation. It also makes it possible to recognise that the ability of NGOs to structure certain state positions is partially a function of a lack of state resources (or interest) but also a function of personal relationships between sets of actors. In my case being able to make these identifications has come about by going to dinner with these actors, listening and learning about the multiple ways in which their lives have connected through space and time; in fact, it has helped me understand how the very existence of the CBD called some of these relationships into being and drew people together.
15. Van Vree's (1999) work focusses exclusively on Europe and North America. This is one of its most useful contributions because it highlights that meetings, at least as they relate to negotiations over international environmental governance, have specific cultural origins and that the international negotiating process contains within it very specific traces of European political history. The practises and protocols that govern international negotiations invoke culturally specific norms revolving around relations of authority that are not at all universal, but have spread over time through processes such as imperialism and capitalism.
16. Reflections of this insight abound in popular culture—for example, the novels of David Lodge or Jason Reitman's 2009 film *Up in the Air*.
17. One of the most ironic scenes during the WCC occurred toward the end of the opening ceremonies when two young environmental protesters appeared on stage with a banner opposing the destruction, through urban sprawl, of a natural park north of Barcelona. They were quickly rustled off stage by security. So, here we had environmental protesters being physically removed from the stage of one of the world's largest gatherings of people who self-identify as leaders of the 'environmental movement'. Following Van Vree (1999), this removal would have occurred not because of their message but because of the way in which it was presented, the breach of meeting protocol; in essence the form of the meeting, and the coherence of the spectacle (defended by force), superceded the substance of the protester's message.
18. Riles's ethnographic study of Fijian bureaucrats and activists (2000) as they prepare for the Fourth World Conference on Women in Beijing provides a richly grounded analysis of the transcendent social life of 'the meeting'. Thanks to Jessica Dempsey for introducing me to Riles's work.

19. In using the word 'association' I am following Latour (2007: 65), who does not use the term to mean some formalised collective of individuals but to signal 'the social', in which 'social is the name of a type of momentary association which is characterised by the way it gathers together into new shapes' facilitated by effort, intent and mechanisms that work to shift weak and hard-to-maintain social ties into more durable kinds of links. It is the creation of these durable links—associations rather than social ties—that we see happening through events like the WCC.
20. Later renamed the Business and Biodiversity Programme.
21. The World Parks Congress is a gathering of international actors and organisations involved in protected area research and management. It is convened by the IUCN's World Commission on Protected Areas (WCPA) once every ten years.
22. Formerly the International Council on Mining and Minerals.
23. For example, to convert the arrangements reached through the fickle personal relationships between often-transient staff members and consultants in different organisations into a process for codified agreements seen to be sanctioned by the organisation's membership.
24. With a registration fee that hovered around €400 for most people, the WCC is very much a middle-class affair. Aside from the need for an institutional affiliation to qualify for admission, much of 'the world' is excluded simply because attendance is beyond their means.
25. In a nod to the location of the WCC, *Aliances*, the Catalan word for alliances, was used in the place of workshop.
26. Notably, only one business-oriented proposal was submitted by a governmental agency.
27. Previously, the private sector had gained access through events organised by units with the IUCN Secretariat. The presence of private sector actors as partners at the WCC also represents the 'taking-of-a-stand' by the IUCN Secretariat in a long-running debate between members and some Secretariat staff over a statute prohibiting private sector organisations from becoming IUCN members. Most of the NGO members have historically opposed private sector membership, while increasing support for it has gained ground among Secretariat staff. According to one former IUCN staff member, 'What I would argue is when people complained about BP, I'd say "Shit, we have the US government as a state member of IUCN so I mean, you know, you know, China's a state member so if we can deal with them, we can deal with anybody." I mean ... you know, there's no angels on this planet'.
28. http://www.youtube.com/watch?v=-aa-MWP2F4o
29. What is most notable in this remark was that 'Future Leaders' were not at the event to learn but to instruct, and to network. What is even more notable is the absence of such support to bring other 'stakeholders' to the meeting to present their perspectives on important issues.

30. http://cmsdata.iucn.org/downloads/markets_journey_english_26_sep_2.pdf
31. In some cases, affiliations were difficult to identify, as a number of assistants were secondees from IUCN to the private sector and vice versa. Notably, de Koning, one of the 'Journey Guides', was identified as an IUCN rather than a Shell employee. de Koning was seconded from Shell to the IUCN in February 2008 for a period of two years with the responsibility to 'build the relationship between Shell and IUCN.' She had three main tasks: 'i) Assist IUCN in improving its business skills through the transfer of appropriate skills from qualified Shell specialists; ii) Assist IUCN in developing capacity to engage more effectively with business on biodiversity issues; and iii) Support Shell businesses in the identification of biodiversity risks to their business and provide help to address such risks through links to IUCN expertise and networks'.
http://cms.iucn.org/about/work/programmes/business/bbp_our_work/bbp_shell/bbp_shellsecondment/. Accessed 8 February 2010.
32. Being in specific places at particular times is extremely important to identifying those markers and the difficulty of covering concurrent sessions at events (e.g., the WCC) highlights the importance of a collaborative approach, as more individuals sharing observations are likely to identify more of those markers.
33. The 'Markets and Business Journey' programme can be accessed online: http://cmsdata.iucn.org/downloads/markets_journey_english_26_sep_2.pdf . Accessed 10 August 2010.
34. As an experiment, I intentionally asked what I anticipated would be two mildly provocative questions in sessions I attended and felt a slight hostility directed at me from both the audience and the panel.
35. It is widely recognised among IUCN members that 'new partners' is code for 'the private sector'.
36. 'Biodiversity is our business' has become a widely used catch phrase among Business and Biodiversity Initiatives.
37. They also are communicating, through a variety of media, their intent to align their organisational environment with that of other organisations and institutions engaged in transnational environmental governance (MacDonald 2008).
38. Portions of this opening ceremony can be seen online: http://www.youtube.com/watch?v=Hs4nhph6Y-k . Accessed 9 February 2010.
39. It was clear that celebrities, such as Ted Turner and Mohammed Yunus, had nothing to say about biodiversity, but that was not their function. They were there to draw attention to the IUCN; to confer authority on the WCC by virtue of their celebrity—much as Royalty represented a sort of consecration—and, perhaps most important, to facilitate the personal

and organisational objectives of the senior IUCN leadership to acquire the kind of access to a transnational capitalist class that can be mediated by figures like Turner, Chair of the UN Foundation, and Yunus, winner of a Nobel Prize.

40. Conveniently forgetting to mention that it always has been 'youth' that have sustained the environmental movement, even in the absence of mobile technology. In fact, a case could be made, based on the presentations made during the plenary that technology is being used as a form of material domination to align the message of 'youth' with that of the (upper) middle-aged folk running large conservation organisations and telecommunications firms, thus robbing 'youth' of their historically transformative and transgressive potential.
41. With an ongoing decline in framework funding, and the lack of a public profile, the IUCN name has become a point of concern for the Secretariat, which has begun to focus heavily on 'brand identity'. It was notable during the Congress that IUCN senior staff made a point of noting that the WCC 'attracted over 400 media people and they released some 5000 or even more stories and video clips and other things around the world. So we actually reach millions of people ... who previously had not realized how important our work is to their everyday life'.
42. In addition, there is little doubt that what Debord (1990) called the 'mediatic status' of figures (e.g., Turner and Yunus) is what the IUCN leadership sought to exploit. Their remarks certainly had nothing to offer in relation to biodiversity conservation. But that does not matter in a world in which the acquisition of that status confers more importance than the value of anything an individual might be capable of doing and where that status is easily transferable to any other domain of knowledge. Listening to Ted Turner, as a keynote speaker at the WCC, sadly does not seem strange in a world characterised by the 'generalized disappearance of all true competence. A financier can be a singer, a lawyer a police spy, a baker can parade his literary tastes, an actor can be president, a chef can philosophise on the movements of baking as if they were landmarks in universal history. Each can join the spectacle, in order publicly to adopt, or sometimes secretly practise, an entirely different activity from whatever specialty first made their name' (http://www.notbored.org/commentaires.html. Accessed 1 June 2001).
43. When studying organisations, it is important never to lose track of the way in which personal aspirations, ambition and desires of organisational leaders shape organisational action, and the degree to which those actions are aligned with the interests of the positions to which those individuals aspire.
44. As simple examples, we might point to the question of various interests weighing in on how to define 'sustainability', which is said to be at the core of the organisation's mandate, or whether the organisation should revert

to the consistent use of 'nature' in the place of 'biodiversity'. Material struggles are much more base and revolve around the allocation of funds between units and commissions within the organisation; or the apparent competition for donor funds between a Secretariat that is largely project-oriented, and most of the NGO members that rely on donor funding for survival.

45. In a rare display of overt activism, the Secretariat actively resisted the motion to terminate its agreement with Shell. This resistance had three primary points: (a) termination would cause the organisation to lose revenue of approximately USD1.3 million; (b) it would expose IUCN to potential legal ramifications; and (c) it would compromise IUCN's ability to secure future partnerships. The remarks of this oil executive, who was not a Shell employee, would seem to confirm the latter belief, at least for the oil sector. But they also reveal a striking lack of understanding of the organisation in which the Secretariat is meant to service the membership; take direction from the membership; and, by statute, cannot prohibit members from submitting motions.
46. Portions of this agreement can be found online but were only uploaded after significant pressure from certain IUCN NGO members: http://www.iucn.org/about/work/programmes/business/bbp_our_work/bbp_shell/
47. But even here, Van Vree's point regarding the constraint of affect and emotion was apparent. In one evening contact group created to refine the motion calling on the IUCN to terminate its agreement with Shell, about 50 people had been abiding by 'the rules', seeking permission from the Chair to speak, identifying themselves before speaking, offering their contributions and yielding the floor. Toward the end of the evening two Australian delegates entered the room and began to offer input that, while seeming to echo the sentiment of the majority of people in the room, contravened the rules that the chair had established at the start of the evening and engaged with the substance rather than the mechanism of the motion. The response was clear. Murmurs of discomfort among the group were summed up by an Australian woman who turned to me and whispered, 'where'd the aggro [aggression] come from all of a sudden?'
48. For example, the IUCNs position in a wider institutional network of neoliberal environmental governance that confers legitimacy and provides material resources (MacDonald 2008).
49. That conservation is configured around more dominant political projects is not a particularly new insight. There is a wealth of work, for example, that situates 'nature' and its conservation relative to projects of imperialism and nationalism.

50. See MacDonald (2006) for an example of the unintended ways in which a focus on 'market-based mechanisms' in the Global Environment Facility affected a small village in northern Pakistan.

References

Barnett, M., and M. Finnemore. 2004. *Rules for the World: International Organizations in Global Politics*. Ithaca: Cornell University Press.

Bassey, N. 2009. Letter dated Amsterdam, The Netherlands, 9 January 2009, from Nnimmo Bassey, Friends of the Earth International Chair, Environmental Rights Action/Friends of the Earth Nigeria Executive Director, to Julia Marton Lefevre, Director General, IUCN. http://www.foei.org/wp-content/uploads/2014/04/IUCN_FoEI-letter-09-jan-09.pdf. Accessed 23 July 2017.

Brechin, S.R., and J. Swanson. 2008. Private Sector Financing of International Biodiversity Conservation: An Exploration of the Funding of Large Conservation NGOs. Paper presented at the workshop, *Conservation and Capitalism*, School of Environment and Development, University of Manchester, Manchester, UK, 8–10 September 2008.

Bruno, K., and J. Karliner. 2002. *Earthsummit.biz: The Corporate Takeover of Sustainable Development*. Oakland: Food First Books.

Dahlén, T. 1997. *Among the Interculturalists*. Stockholm: Almqvist and Wiksell International.

Debord, G. 1967. *The Society of the Spectacle*. Tr. 1977. London: Rebel Press (2004).

———. 1990. *Comments on The Society of the Spectacle*. New York: Verso.

Elbers, W. 2004. *Doing Business with Business: Development NGOs Interacting with the Corporate Sector*. Nijmegen: Centre for International Issues, Radboud University.

Fox, D.J. 1998. *An Ethnography of Four Non-governmental Organizations*. Lewiston: The Edwin Mellen Press.

Frynas, J.G. 2005. The False Developmental Promise of Corporate Social Responsibility: Evidence from Multinational Oil Companies. *International Affairs* 81 (3): 581–598.

Haas, P. 1992. Epistemic Communities and International Policy Coordination. *International Organization* 46: 1–35.

Hajer, M.A. 2006. The Living Institutions of the EU: Analyzing Governance as Performance. *Perspectives on European Politics and Society* 7: 41–55.

Hajer, M.A., and W. Versteeg. 2005. Performing Governance Through Networks. *European Political Science* 4: 340–347.

Hannerz, U. 2003. Being There ... and There ... and There!: Reflections on Multi-site Ethnography. *Ethnography* 4: 201–216.

Harper, R.H.R. 1998. *Inside the IMF: An Ethnography of Documents, Technology and Organizational Action*. San Diego: Academic Press.

Hutton, J., and B. Dickson. 2000. *Endangered Species: Threatened Convention: The Past, Present and Future of CITES, the Convention on International Trade in Endangered Species of Wild Fauna and Flora*. London: Earthscan.

Igoe, J., K. Neves, and D. Brockington. 2010. A Spectacular Eco-tour Around the Historic Bloc: Theorising the Convergence of Biodiversity Conservation and Capitalist Expansion. *Antipode* 42 (3): 486–512.

Kean, K. 2001. Following the Thread in Computer Conferences. *Computers & Education* 37: 81–99.

Lahsen, M. 2007. Anthropology and the Trouble of Risk Society. *Anthropology News* 48 (9): 9–10.

Latour, B. 2004. *Politics of Nature: How to Bring the Sciences into Democracy*. Cambridge, MA: Harvard University Press.

———. 2007. *Reassembling the Social: An Introduction to Actor-Network Theory*. Oxford: Oxford University Press.

Latour, B., and S. Woolgar. 1986. *Laboratory Life: The Social Construction of Scientific Facts*. Princeton: Princeton University Press.

Levy, D.L. 2005. Business and the Evolution of the Climate Regime: The Dynamics of Corporate Strategies. In *The Business of Global Environmental Governance*, ed. D.L. Levy and P. Newell, 73–104. Cambridge: MIT Press.

Lewis, D. 2003. NGOs, Organizational Culture, and Institutional Sustainability. *The Annals of the American Academy of Political and Social Science* 590: 212–226.

Litfin, K.T. 1994. *Ozone Discourses: Science and Politics in Global Environmental Cooperation*. New York: Columbia University Press.

Little, P.E. 1995. Ritual, Power and Ethnography at the Rio Earth Summit. *Critique of Anthropology* 15: 265–288.

MacAloon, J.J. 1992. Sport, Science, and Intercultural Relations: Reflections on Recent Trends in Olympic Scientific Meetings. *Olympika: The International Journal of Olympic Studies* 1: 1–28.

MacDonald, K.I. 2003. IUCN–The World Conservation Union: A History of Constraint. Invited Lecture to the Permanent Workshop of the Centre for Philosophy of Law, Higher Institute for Philosophy, Catholic Univeristy of Louvain (UCL), Louvain-la-neuve, Belgium, 6 February 2003. http://perso.cpdr.ucl.ac.be/maesschalck/MacDonaldInstitutional_Reflexivity_and_IUCN-17.02.03.pdf. Accessed 28 May 2010.

———. 2006. Global Hunting Grounds: Power, Scale and Ecology in the Negotiation of Conservation. *Cultural Geographies* 12: 259–291.

———. 2008. The Devil Is in the (Bio)diversity: Neoliberalism and the Restructuring of Conservation. Paper presented at the workshop, *Conservation and Capitalism*, School of Environment and Development, University of Manchester, Manchester, UK, 8–10 September 2008. https://tspace.library.utoronto.ca/handle/1807/24408. Accessed 28 May 2009.

———. 2010. The Devil Is in the (Bio)diversity: Private Sector 'Engagement' and the Restructuring of Biodiversity Conservation. *Antipode* 42: 513–550.

MacKenzie, D. 2009. *Material Markets: How Economic Agents Are Constructed.* Oxford: Oxford University Press.

Markowitz, L.B. 2001. Finding the Field: Notes on the Ethnography of NGOs. *Human Organization* 6: 40–46.

Mosse, D. 2001. Social Research in Rural Development Projects. In *Inside Organizations: Anthropologists at Work*, ed. D.N. Gellner and E. Hirsch, 159–181. Oxford: Berg.

Nader, L. 1972. Up the Anthropologist: Perspectives Gained from Studying Up. In *Reinventing Anthropology*, ed. D.H. Hymes, 284–311. New York: Pantheon Books.

———. 1995. Civilization and Its Negotiators. In *Understanding Disputes: The Politics of Argument*, ed. P. Caplan, 39–63. Oxford: Berg.

Poncelet, E.C. 1990. An Ethnographic Approach to the Conference: An Anthropological Perspective. In *Japanese/American Technological Innovation: The Influence of Cultural Differences on Japanese and American Innovation in Advanced Materials: Proceedings of the Symposium on Japanese/American Technological Innovation*, ed. W.D. Kingery, 259–262. Tucson: University of Arizona.

Poncelot, E. 2004. *Partnering for the Environment: Multistakeholder Collaboration in a Changing World.* Lanham: Rowman & Littlefield.

Pulman-Jones, S. 2001. Observing Other Observers: Anthropological Fieldwork in a Unit for Children with Chronic Emotional and Behavioural Problems. In *Inside Organizations: Anthropologists at Work*, ed. D.N. Gelner and E. Hirsch, 117–136. Oxford: Berg.

Reed, M. 2001. From Eating to Meeting: The Rise of Meeting Society. *Theory, Culture and Society* 18 (5): 131–143.

Riles, A. 2000. *The Network Inside Out.* Ann Arbor: University of Michigan Press.

Rosen, M. 1991. Coming to Terms with the Field: Understanding and Doing Organizational Ethnography. *Journal of Management Studies* 28: 1–24.

Van Vree, W. 1999. *Meetings, Manners and Civilization: The Development of Modern Meeting Behaviour.* Leicester: Leicester University Press.

Vogler, J. 2003. Taking Institutions Seriously: How Regime Analysis Can Be Relevant to Multilevel Environmental Governance. *Global Environmental Politics* 3: 25–39.

Young, O.R. 1982. *Resource Regimes: Natural Resources and Social Institutions.* Berkeley: University of California Press.

———. 1989. *International Cooperation: Building Regimes for Natural Resources and the Environment.* Ithaca: Cornell University Press.

———. 1994. *International Governance: Protecting the Environment in a Stateless Society.* Ithaca: Cornell University Press.

CHAPTER 5

The Strategies and Effectiveness of Conservation NGOs in the Global Voluntary Standards: The Case of the Roundtable on Sustainable Palm Oil

Denis Ruysschaert and Denis Salles

Introduction

Tropical forest biodiversity is declining at an alarming rate of more than 1% per year, largely because of agricultural expansion into forests (WWF et al. 2014). Southeast Asian countries, especially Indonesia and Malaysia, follow this trend because of the recent surge of the palm oil industries that are subsidised by their respective governments (Pye and Bhattacharya 2013; USDA 2014). Once a marginal crop in the 1980s, large-scale oil palm plantations are transforming the low land forest landscape in both the countries. In Indonesia alone, plantations cover 110 thousand square kilometres

D. Ruysschaert (✉)
Centre d'Étude et de Recherche Travail, Pouvoir (CERTOP),
Université Toulouse, Toulouse, France

D. Salles
Unité Environnement, territoires et infrastructures (ETBX),
IRSTEA, Bordeaux, Cestas, France

P.B. Larsen, D. Brockington (eds.), *The Anthropology of Conservation NGOs*, Palgrave Studies in Anthropology of Sustainability, DOI 10.1007/978-3-319-60579-1_5

(sq. km), including 60% for large-scale plantations of several thousand hectares each (DGEC 2014). An additional 150 to 270 thousand sq. km are under permits for further oil palm plantations (Colchester and Chao 2011). Peat land forest areas in both countries are particularly threatened. In 2010, plantations covered more than 31,000 sq. km of peat land, with projections of between 60,000 and 90,000 sq. km by 2020 (Miettinen et al. 2012).

Large-scale oil palm plantations greatly impact wild species. The plantations host 65% less biodiversity than natural forests. They also contribute to forest fragmentation and create an ecological barrier for species (Fitzhebert et al. 2008). Mammals are particularly vulnerable because they require large territories of several square kilometres for a viable population. Mainly because of recent oil palm expansion, orangutan, tiger, elephant and rhinoceros in Southeast Asia have been classified as 'endangered' or 'critically endangered'on the IUCN Red List (IUCN 2015). For instance, the critically endangered Sumatran Orangutan (*Pongo abelli*), with a current population of 6600 and a habitat of around 8000 sq. km, has lost an estimated 90% of its original habitat during the twentieth century. This species still is losing about 50 sq. km of its forest habitat each year mainly as a result of oil palm plantations (Wich et al. 2011; Ruysschaert and Salles 2014). The rise of palm oil production also resulted in 3500 land conflicts about palm plantations with local communities in Indonesia between 1997 and 2009 (Jiwan 2013).

It is within this context that the World Wide Fund for Nature (WWF) promoted the establishment of the multi-stakeholder private association, the Roundtable on Sustainable Palm Oil (RSPO), the aim of which is to produce socially and environmental responsible palm oil. In 2001, the WWF approached Western firms with a history of operating in Southeast Asia, primarily those based in the United Kingdom and the Netherlands (RSPO 2002). Banks and agrobusiness firms in those countries were seeking to secure their long-term supply and to protect themselves from possible negative environmental campaigns (RSPO 2002). Indeed, their major financial institutions (e.g., Rabobank, Standard Chartered) were financing the firms converting forest into large-scale oil palm plantations. Their main agrobusiness firms were also buying large quantities of palm oil from Southeast Asia (RSPO 2002). This included processors (e.g., AarhusKarlshamn, AAK), consumer goods' manufacturers (e.g., Unilever) and retailers (e.g., Mark and Spencer and Sainsbury's). As a prominent example, Unilever is the biggest single user of palm oil in the world.

In 2004, the RSPO was established as an association comprised of all private stakeholders within the palm oil supply chain (RSPO 2004). It was divided into seven membership categories: (1) palm oil growers, (2) palm oil processors, (3) consumer goods' manufacturers, (4) environmental NGOs, (5) social NGOs, (6) banks and/or investors, and (7) retailers. In 2007, the members agreed on a standard of production, including social and environmental principles and criteria (RSPO 2007). As the first global standard in agriculture in the tropics, it has been replicated for other commodities such as aquaculture, agrofuels, cotton, sugar cane and soy.[1] As of June 2015, the RSPO has more than 1,200 ordinary members (RSPO 2015a) and certifies 20% of the globally produced palm oil as sustainable (RSPO 2015b).

Despite the rising number of RSPO members and the volume produced according the standard, lowland natural forest loss is still increasing in Indonesia, with 84,000 sq. km in 2012. Between 2001 and 2012, Indonesia lost an average 56,000 sq. km of forest per year. Of this loss 40% is attributable to large-scale oil palm plantations on peatland (Margono et al. 2014). Overall, the RSPO has not been effective in curbing deforestation because of oil palm plantations (McCarthy 2012; Ruysschaert and Salles 2014; Ruysschaert 2016).

At the same time, the number of conservation NGO members joining the RSPO has continuously risen, with a total of 30 conservation NGO members at the end of June 2015 (RSPO 2015a). Therefore, this study seeks to understand this growing contradiction between an apparently ineffective RSPO and the continuous, and even increasing, interest of conservation NGOs in this scheme. By understanding the roles and effectiveness of conservation NGOs with respect to the RSPO standard, the research seeks to make sense of this inconsistency.

This chapter takes the position of considering the diversity of conservation NGOs engaged with the RSPO. It formulates the hypothesis that NGOs are strategic players with clear conservation goals that mobilise their scarce resources to achieve these goals (Friedberg 1991). Based on what the resources and goals are, NGOs adopt several different roles. NGOs can mobilise four broad types of resources to influence the terms of the relationships to their advantage: (1) expertise, (2) affiliations outside the system (e.g., media, government), (3) communication and information, and (4) institutional rules within the system (Crozier and Frieberg 1977).

The RSPO's construction could limit the NGOs' ability to have an impact in three complementary ways. The initial reason would be that the institutionalisation forces each individual NGO to adopt a specific strategy as their resources are decreasing while RSPO is structuring.

Then, the institutionalisation may impede collaboration between NGOs using various strategies, whereas this collaboration in fact would be a key to getting leverage. Finally, some NGOs may be structurally excluded because of their incompatibility with the prevalent capitalistic discourse within the RSPO, although their participation is fundamental, both in terms of long-term conservation gains and social justice (Fouilleux 2013; Pye 2013; Cheyns 2014; Ruysschaert 2016).

The Importance of the NGO Effectiveness Question to Research

So far researchers have demonstrated that the emergence of certifications for global commodities, such as those of the RSPO, can be qualified as a new capitalist instrument (Fouilleux and Goulet 2012), promoted by dominant economic players (Fouilleux 2013; Oosterveer 2014) and major NGOs (Cheyns 2012; Escobar and Cheyns 2012; Pye 2013), and operating at the expense of local people (Ponte et al. 2011; Fouilleux 2013; Cheyns 2014).

These researchers meet the growing political ecologists' consensus on synergistic relationships between the protection of nature and neoliberalism, where the largest conservation NGOs in terms of staff and turn over—for example, the WWF or Conservation International (CI)—would directly support market expansion (Igoe and Brockington 2007; Benjaminsen and Svarstad 2010; Igoe et al. 2010a, b; Büscher et al. 2014). The main reason being that the conservation NGOs and capitalist agendas are driven by the convergence of networks of interests (Igoe and Brockington 2007; Barker 2009; Brockington and Duffy 2010; Igoe et al. 2010b). This limits the NGOs' effectiveness because they tend to overlook local communities (Chaplin 2004; Igoe and Brockington 2007; Tumusiime and Svarstad 2011) while maintaining that cultural diversity is inseparable from biodiversity conservation (Igoe 2005; Peluso 2012).

Why the NGOs are behaving in such a negative manner from a conservation perspective remains a subject for discussion. One reason seems to be the need for fresh financial resources, as these large NGOs are ever-expanding in terms of budget and human resources (Chaplin 2004; Barker 2009; Sachedina et al. 2010). As such, the NGOs tend to behave as a collection of projects without any clear overarching purpose because of a lack of accountability mechanisms (Sachedina et al. 2010) or, worse, being enslaved to their donors' priorities (Chaplin 2004; Barker 2009) and their neoliberal agenda (Brockington and Duffy 2010).

If these findings are to be relevant for the biggest NGOs, it tends to overlook the whole range of conservation NGOs (Brockington 2011), even though NGOs' performance in reaching their conservation goals can vary considerably depending on the NGO and the region in which it operates (Igoe et al. 2010b). As a main consequence, studies tend to fall short of solutions aside from the rather populist recommendations to decentralize conservation and its finance to the locals (McCarthy 2002; Barker 2009) and join the fight against neoliberalism (Barker 2009; Sachedina et al. 2010). Despite that, what can NGOs do to better achieve conservation outcomes?

Methodology

To analyse the roles and effectiveness of conservation NGOs in the RSPO, this research takes a political ecology approach. In other words, it is the conjunction of a pattern of interactions made of economic interests (i.e., expansion of oil palm plantations for agribusiness), ecological changes (i.e., destruction of orangutan rainforest habitat), and political battles (i.e., designation of land use and deprivation of local communities of their territories) (Gautier and Benjaminsen 2012). As such, the study is firmly rooted in an acute conflict on land use allocation between conservation NGOs and oil palm growers because of orangutan habitat conversion into plantations in the lowland forests of Malaysia and Indonesia. This conflict between conservationists and growers takes place against the background of the broader issue of denial of forested land of local communities in those countries (Peluso 2012). This approach has the advantage of limiting the scope of the study and maintaining a common thread from the global context of the RSPO to the local tangible ecological reality and social conflict in Southeast Asia.

This study involves access to a wide range of materials. The research benefits from in situ observation of the NGOs engaging with the RSPO. Indeed, from October 2006 to October 2011, one of the authors was a staff member of a conservation NGO that is an RSPO member. This provided direct access to information on the specific objectives and the resources of the NGOs involved during the RSPO annual meetings, General Assembly (GA) and RSPO bodies linked to biodiversity conservation—for example, the Biodiversity and High Conservation Values Working Group (BHCV WG). This provided firsthand information to detail the 'playing field', which is how the RSPO functions, and categories (or roles or engagement regimes or strategies) of NGOs based on the resources they mobilise and their conservation goals.

The effectiveness and structural constraints of each of these engagement regimes were investigated by conducting 49 semistructured interviews from November 2011 to July 2013, which represented 33 institutions, including 11 NGOs—that is, Birdlife, the CI, PanEco, Greenpeace Indonesia, Greenpeace International, Leuser International Foundation, the Sumatran Orangutan Society (SOS), the World Conservation Society (WCS), the WWF, Yayasan Ekosistem Lestari and Yayasan Pulau Banyak—dealing with palm oil. In addition, Internet-based research was undertaken to obtain publicly available information (e.g., reports, press releases and web pages) between 2001 and June 2015 on the RSPO and on all the main conservation NGOs engaged with it. This research was complemented by field observations during the European RSPO meeting in London in June 2014 and the RSPO annual meeting in November 2014.

Argument

The argument is developed in three parts. First, it describes the 'playing field'—how the RSPO functions formally as an institution. Second, it details the four strategies (or roles) of NGOs based on the resources they mobilise and their conservation goals: the collaborative, the opposing, the opportunistic, and the sceptics. Third, it explains the relationship of each of these NGO strategies with the RSPO institutionalising, and how it limits their impact.

How the RSPO Is Formally Functioning

The RSPO's objective is to promote the growth and use of sustainable palm oil (RSPO 2004). Sustainability application is defined in the 50-page guidance document called *Principles and Criteria for the Production of Sustainable Palm Oil*; it details eight principles,[2] along with the associated criteria and indicators—that is, 5.2 and 7.3, which specifically deal with biodiversity conservation. Criterion 5.2 requires growers to conserve rare species, habitats and control hunting (RSPO 2013: 27). Criterion 7.3 requires that new palm plantings starting from November 2005 onward do not replace primary forest or high conservation value (HCV) areas. The HCV areas are designated as such because of their importance for biodiversity conservation or for the local community's well-being (RSPO 2013: 50).

The choice of criteria and its related indicators resulted from extensive negotiations among members with respect to three fundamental rules:

(1) inclusive participation from each member category, (2) consensus-building in reaching agreements, and (3) transparency during the negotiation process and the decision making. In addition to these rules, debates among members were structured by a scientific research-based managerial discourse so as to depoliticise any deliberation among them (Cheyns 2012).

Approved at the RSPO General Assembly (GA) of 2007 (RSPO 2007) the guidance document was revised in 2013 (RSPO 2013) to strengthen its environmental criteria and indicators. Sustainability, therefore, has been an evolving concept in which each member category defends its own interests. Following the approval of the guidance document, the RSPO introduced the category of certified sustainable palm oil (CSPO) to the market in 2008. This enabled downstream firms to label the final product with a distinctive CSPO trademark.

The RSPO is a system formed with three distinct governing bodies: the GA, the Board of Governors (BG) and the Secretariat. In between the GAs, the BG provides strategic and operational direction. The BG is comprised of 16 members, 4 growers and 2 members each from other 7 membership categories. The BG members are elected from their specific categories during the GA for a two-year term, and there is great stability in those BG roles. Since the beginning, a manufacturer of consumer goods, Unilever, is the president, the processor AAK-UK is the treasurer, and the environmental NGO WWF is the vice president.

The GA is an annual meeting where members can propose new resolutions. Resolutions often seek to interpret the implementation of the guidance document to favour a specific interest, which, for the environmental NGOs, is conservation. The GA decision-making process is undertaken by casting votes and making endorsements by a simple majority. As a result, to pass any resolution, a member must garner support from downstream firms (i.e., palm oil processors, consumer goods' manufacturers and retailers) because they represent about 80% of the RSPO membership (RSPO 2015a).

To implement these decisions, the BG establishes working groups or task forces; they are made up of members and function according to RSPO core principles. Working groups provide recommendations to the BG on how to implement the GA decisions. With a growing number of working groups, the RSPO has adopted a formal operational structure, which consists of four permanent standing committees: Standards and Certification, Trade and Traceability, Communication and Claims and Finance. Under each of these committees, the working groups deal with long-term issues and task forces deal with short-term issues.

The Secretariat manages the logistical aspects of the RSPO, organises the yearly roundtable meetings associated with the GA, promotes the RSPO worldwide, facilitates the work of the standing committees, and implements the GA's decisions under BG guidance.

The Various Strategies of the NGOs

This study found that conservation NGOs adopt four main forms of engagement with the RSPO based on two main criteria: the resources they are allocated and the specific objectives they pursue. These categories are: collaborative, opposing, opportunistic and sceptic. The criteria to develop each of the categories are detailed in Table 5.1.

Collaborative NGOs' Strategies and Their Limitations

Collaborative NGOs are RSPO members aiming to protect tropical forests through collaboration with the business sector. Usually, these are powerful international Western NGOs with headquarters in Europe (e.g., the WWF, Zoological Society of London, Wetlands International) or the United States (e.g., CI, the World Resources Institute). They abide by the RSPO's overall vision to 'transform markets to make sustainable palm oil the norm' (RSPO 2012: 8). This vision is similar to the WWF's Market Transformation Initiative Strategy, which showcases how NGOs can influence the overall communication strategy of the RSPO (WWF 2012).

This type of NGO invests a considerable amount of human and financial resources, working within the RSPO system to reform the oil palm sector. The programme director of one of these international NGOs mentioned that 'we have a number of people working full time on palm oil in Singapore on finance, in Indonesia and Malaysia on producers, in China and India, and across Europe and in the US on [pursuing] ... business and industry engagement'.[3]

These resources are channelised toward four broad lines of actions to influence the RSPO system and reach conservation goals. First, they aim to create rules in the RSPO that support their conservation goals. In practise, this means putting forward the RSPO GA's decisions that force growers to implement the guidance document to the benefit of conservation (see section later about analysing of GA decisions regarding biodiversity and greenhouse gases). Even though conservation NGOs makeup less than 3% (or 30 individual members) of all RSPO members, the RSPO GA adopted most of their decisions because of the support from several downstream firms. These

Table 5.1 Typology of engagements of environmental NGOs with the RSPO

Form of engagement	*Examples of conservation NGOs*	*Resources allocated to the RSPO*	*Main objective pursued with the RSPO*
Collaborative	Generalist on environment: Conservation International, World Resource Institute, WWF	RSPO members: • Actively participate in it. Executive Board and various commissions • Engage systemically within it	Reform the palm oil sector by influencing the RSPO from within
Opposing	Generalist on environment: Friends of the Earth, Greenpeace	Not RSPO members: • Explain files against RSPO growers • Selected strategically • Use the RSPO as a forum to protest	Radically criticise RSPO members, and therefore RSPO in its legitimacy to reform the palm oil sector from within Seek to get a global moratorium on deforestation and reform the palm oil sector
Opportunistic	Organisation focussing on orangutans: Borneo Orangutan Survival, Hutan, PanEco and Sumatran Orangutan Society	RSPO members: • Exclusively participate to annual meetings and GAs • Use the RSPO as a tribune	Get specific decisions to protect specific forests to save various habitats for orangutans (Tripa, Bukit Tigah Puluh, degraded forest)
Sceptic	Generalist on environment with human rights-based approach: Most environment and social associations in the producing countries—Forest Peoples Programme, Sawit Watch	Usually not RSPO members: • No resources allocated • No participation If RSPO member, participate on the BG and commissions, but participation decreases over time	Seek community rights recognition, question the production model Overtime, tend to avoid links with the RSPO because of their inability to influence it. RSPO tends to become secondary in their strategy, distrusting this initiative that diverts resources that could be better used elsewhere

firms support the NGOs' decisions not only because they reduce reputation risks but also because they do not bear the cost of implementing decisions that are supported by growers alone.

Second, they take strategic positions within the RSPO. They are a part of the RSPO GB (i.e., CI, then World Resources Institute and WWF) to influence RSPO functioning. They also participate in the working groups and task forces. In those working groups, they hold strategic positions as cochairs—for example, the World Resources Institute for Biodiversity and Global Environmental Centre for the working group on peat. This internal lobbying by collaborative NGOs is often overlooked because such political manoeuvring happens far from the focus of public attention. As explained by a former BG member: 'WWF being inside the organisation was very critical, as I perceived them when I was a member of the board. They were not at all not criticising the companies, … but are very critical inside the organisation, pushing for criteria'.[4]

Third, the NGOs commonly use scientific facts. Because collaborative conservation NGOs are made up of scientists, they value the science-based arguments within the RSPO. As an example, the Zoological Society of London (ZSL) states that it is important 'to ensure that emerging policies and good practice guidelines aiming to promote sustainable production are based on the best possible science' (Persey et al. 2011: 6). As such, they provide scientific expertise on topics such as biodiversity (e.g., ZSL), satellite mapping (e.g., World Resources Institute) and greenhouse gases (e.g., CI and Wetlands International).

Fourth, NGOs widely campaign to raise public awareness, creating external pressure for RSPO members to fulfil their commitments. To avoid name and shame tactics, they use collaborative techniques consisting of benchmarking diverse RSPO members, comparing commitments and encouraging them to get certified (WWF 2013). As developed later, the mobilisation of these four resources has been an effective technique to challenge RSPO norms, but it also has led to some structural constraints.

Limitation of the Collaborative Approach to Conserve Biodiversity Areas

The RSPO has developed the HCV concept to conserve areas the most important to biodiversity or human well-being. Its definition and application are subject to intense discussion between growers and conservationists that have opposite opinions regarding land use. For growers, conservation areas are thought to be underutilised profit spaces. They seek to limit the size of required conserved areas within their concessions. In contrast, the NGOs consider low land areas as rich in exceptional tropical biodiversity. According to this interpretation, oil palm plantation extensions must take place in already degraded areas because 'there is enough non-forested land suitable for plantation development to allow large increases in production' (Fitzherbert et al. 2008: 545; Persey et al. 2011; Ruysschaert et al. 2011). Conservationists tend to maximise the designated conservation areas well beyond primary forests, including degraded land (RSPO 2010).

To implement the HCV concept, in other words to define the criteria and indicators attached to it, the RSPO Biodiversity and Technical Committee (BTC) was created in 2009. It quickly became clear, however, that this committee had significant scientific information gaps on tropical biology. A representative from this committee summarises the situation: '[S]ometimes, it is not clear at all. Frankly, why is biodiversity important? It is not at all obvious to demonstrate. A lot of science is needed'.[5] Indeed, research remains inconclusive about tropical biodiversity conservation. Results depend on the species and the level of organisation (e.g., genetic, species or ecosystem) studied. Some reports show the importance of fragmented forests, others of the continuous forest (Benedick et al. 2006; Struebig et al. 2008; Edwards et al. 2010; Struebig et al. 2011).

As a result, this temporary committee was institutionalised as the Biodiversity and HCV (BHCV) working groups. Its aim now is 'to provide strategic and technical support' to the RSPO (RSPO 2015f: 1), underlining the lasting nature of knowledge gaps. It met 25 times between April 2009 and January 2015. The BHCV's formal structure is comprised of 12 RSPO members. Two environmental NGOs hold seats in the group, one of them being cochair (ZSL, then World Resources Institute), and a grower holds the other seat. In addition, many other NGO members participate as observers at each meeting, accounting for more than a quarter of the participants (10–20).

The institutionalisation of the BTC into a permanent working group reflects the ongoing and irreducible tensions between conservationists and growers. There are constant information gaps in science and the illusion that full scientific knowledge could arbitrate the divergent interests is sustained. As a result, on one hand, collaborative NGOs seek to consolidate information to support conservation; on the other hand, growers decide which areas can be conserved on their concessions in a pragmatic manner, stating that full knowledge is not yet available. Paid by the grower, evaluators tend to designate the HCV areas in a 'subjective' and 'arbitrary' manner (HCV 2007: 2), reducing the areas to conserve to the most valuable biodiversity areas. In particular, this dependency can lead the auditors to omit secondary or degraded areas because '... the value and performance of these kinds of forest have little meaning in terms of wildlife and [the] environment' (RSPO 2010: 16). This is the type of reasoning that the Sumatran Orangutan Society and the grower PT Sisirau opposes. As explained by an SOS's representative, 'this member was clearing orangutan habitat, ... but claimed it wasn't good quality forest'.[6]

Limitation of the Collaborative Approach to Conserving Primary Forests

With the structural knowledge gap on biodiversity impeding conservation NGOs to mobilise scientific information effectively, they turned their attention to the implementation of Criterion 7.3. This criterion requires preserving primary forest or HCV within the plantations after November 2005. The criterion reduces growers' room for interpretation because it gives a clear-cut end date and specifies which type of forest to conserve.

Nevertheless, primary forest areas are difficult to delineate in practise. In addition, the 2005 date was subject to interpretation. Indeed, the criterion was agreed to at the RSPO GA in 2007 as part of the overall agreement on the guidance document for members who seek to certify plantations. Growers consider that date relevant only when certification is foreseen. They argue that the RSPO founding members were not requested to follow these requirements in the beginning. They also add that some RSPO growers could have taken over other non-RSPO growers who may have clear-cut primary forest.

As one grower explained: 'We had understood that this 2005 rule applies to existing plantations, when we were starting something, we are responsible from the moment where we manage the plantation, not before'.[7] To clarify these points, the RSPO GA decision endorsed the WWF's proposal of 'New Planting Procedures' in 2008, jointly tabled with New Britain Palm Oil, one of the largest oil palm growers in the world.[8] The decision requested growers to preserve primary forest regardless of whether it is seeking certification or not. It came into force in January 2010, despite growers attempts at the 2009 GA to postpone it on the grounds of practical feasibility.

To consider growers' liability, the RSPO BG set up a Compensation Task Force in 2010 as a subunit of the BHCV WG for an initial period of one year, with the aim of providing guidance to the Board of Governors. Four years later, in May 2014, the task force provided a 'voluntary' guidance document

(RSPO 2014: 12). This document seeks the 'continuous improvement' of each grower's particular situation, refusing to expel or suspend them (RSPO 2014: 1).

At these meetings, the NGOs put forward arguments based on scientific evidence. They utilised satellite technology as a method of measuring destroyed primary forests or those of HCV. Still, NGOs were unable to impose their findings for two main reasons. Science-based decisions continue to be problematic. A grower at this task force meeting summarises the issue by stating: 'There is still a lot of work. Coming up with a satellite imagery analysis method that can be implemented and credible is difficult'.[9]

Equally important, the negotiating rules (e.g., inclusive participation and consensus-building) must be respected, which means that the application needs to be for all growers and adapted to each. This in turn has two implications: growers' liability should increase overtime, reflecting the changing rules within the RSPO; and the mechanism needs to give enough incentive for the grower to participate. In other words, the grower would take actions to rehabilitate the forest and not to clear-cut it and pay for its destruction.

The discussions over Criterion 7.3 demonstrate irreconcilable views. For the growers, 'November 2005 is absolute rubbish in the long-term and for the moment it is carved in stone'[10]; for the conservation NGOs, this date was indeed a nonnegotiable. Conservation NGOs have heavily invested in task force meetings by providing scientific evidence and participating physically. They seek to establish an adequate compensation mechanism that would definitively forbid primary forest conversion after 2005. On the opposite side, few growers have participated, blocking advance on the principles fundamental to RSPO's functioning. This leads to the paradoxical situation, in which primary forests can still be cut, even if their preservation is mentioned in the RSPO guidance document.

Limitations of the Collaborative Approach to Mobilise Climate Arguments

Experiencing difficulty mobilising scientific arguments to preserve biodiversity and primary forests, the NGOs sought to protect biodiversity by advancing the climate change agenda. They have been able to do so because of the massive release of greenhouse gases (GHG) from the trees and peat when establishing oil palm plantations within tropical forests. Destruction of these forests to establish plantations releases massive amounts of greenhouse gases. The first research on greenhouse gas emissions from peat in Indonesia revealed striking figures: Indonesia is third in greenhouse gas emissions globally, behind China and the United States (Hooijer et al. 2006). An RSPO member explains that the climate agenda is a powerful argument to save peat land areas from destruction: 'When you hear about the greenhouse gas emissions' study by Deltares, this is an exemplary study, in my view. This proves by A to B that we really cannot do anything on peat lands'.[11]

The large amount of GHG comes from two sources: the trees and the peat. Regarding the trees, the conversion of tropical natural forests into oil palm plantations releases carbon dioxide because tropical forests store above ground around 190,000 kg of carbon per hectare, while oil palm plantations store only around 40,000 kg per hectare; this means that the difference, or about 150,000 kg of carbon per hectare, is released as GHG when natural tropical forest is converted into an oil palm plantation. Even more important is the carbon dioxide released from the peat. This comes from the fires to prepare the land and the oxidation from the dry peat resulting from establishing the necessary drainage channels for the plantations (Agus et al. 2013).

At the RSPO GA in 2008, Wetlands International put forward the decision 'A Moratorium on Palm Oil from Tropical Peat Lands' that requested the 'cessation on any further development of palm oil on tropical peat lands/ (Wetlands International 2008: 1). The RSPO BG refused this submission arguing that it had been modified at the last minute after the due date for submission. It also prevented the submission of the original decision saying that: 'The RSPO [should] adopt that palm oil produced on peat lands be henceforth considered unsustainable until proven otherwise' (RSPO 2008: 33). Using procedural rules, the RSPO BG had found a way to avoid clashes between growers and conservation NGOs.

Instead of the decision, the RSPO BG put forward the core collaborative rules within the RSPO and established the Working Group on Greenhouse Gases. The working group, composed of the different members' categories, had the goal of finding a practical means to reduce GHG emissions by the next GA in November 2009. Facilitated by a Dutch consultant, the group was not able to conclude its work because there was no agreement on greenhouse gas emissions, and because the process was felt to be a kind of neocolonialism.

During the next RSPO GA in 2009, Wetlands International put forward a new proposal, the 'Establishment of a Working Group to Provide Recommendations on How to Deal with Existing Plantations on Peat Lands'. Growers were initially against it. The representative from the Malaysian Palm Oil Association felt that 'the resolution is just to get growers to stop planting oil palm on peat after one or two cycles [which is 25 or 50 years]' (RSPO 2010: 27). Only after Wetlands International ascertained that this would not be the case, a large majority of the growers adopted the decision.[12]

To implement the decision, the RSPO BG established Working Group II on Greenhouse Gases, which met six times between March 2010 and May 2011. It is composed of 30 members (12 from the Executive Board and 18 other RSPO members) and is cochaired by an environmental NGO (Conservation International, then the WWF) and one grower, the Malaysian Palm Oil Association (MPOA), then Wilmar. On paper, this body brings together all the RSPO membership categories. In reality, environmental NGOs (i.e., CI, the Global Environmental Centre, Wetlands International and the WWF) and growers[13] from Indonesia and Malaysia dominate the working group. It established the Working Group on Peat to specifically deal with the tremendous impact of peat on oil palm plantation. The working group is cochaired by an environmental NGO (Global Environment Centre) and a research institution (Indonesian Palm Oil Commission) linked to the Indonesian Ministry of Agriculture. It gathered six growers from Indonesia and Malaysia,[14] five environmental NGOs,[15] four research centres,[16] the HBSC Bank, and the Finnish agrofuel company, Neste Oil.

Working Groups I and II on Greenhouse Gases and the Working Group on Peat have been subject to intense debate between conservation NGOs and growers. This reflects their fierce opposition to each other's conservation goals. Conservation NGOs advanced scientific published research on greenhouse gases. Growers challenged them on the grounds that they are generalities, far from describing the specific realities at the plantation level. Indeed, at the local level, GHG emissions depend on at three parameters: drainage canal depth, number of years of operation and fertiliser dosage.

As one scientist summarised: 'We find a relationship with the depth of the canal and the carbon emission, but it changes over time. In the first few years you have high emissions, and with time it settles on different relationships. In the long term, it settles with low emission with a peat depth of 40 cm or lower'.[17] To fill this scientific knowledge gap on GHG emissions linked to oil palm, NGOs, especially Conservation International, have undertaken some research from the RSPO. In 2013, the RSPO published its findings. But, they are not authoritative as some results are contradictory as a result of the various methodologies used (Killeen and Goon 2013). Therefore, the conservation NGO leading the publication has put out a call for more research instead of advocating for immediate action (Killeen and Goon 2013).

Additionally, growers argued that management principles at the core of RSPO functioning must be respected. Each grower needs to establish its GHG reduction initiatives on a voluntary basis to ensure long-term economic viability of the plantation. Indeed, each grower faces specific circumstances—that is, the state of development of its 25-year plantation, the prospect to extend it onto new peat land areas, the relative area of the plantation on peat compared to the total oil palm plantation area, and the plan for its conservation efforts.

After four years of negotiations, during their April 2013 GA, the RSPO agreed on a new guidance document that takes into account a specific focus on GHG. The document mentions 'to reduce' (RSPO 2013: 32, Criterion 5.6) and 'to minimise' (RSPO 2013: 56, Criterion 7.8) greenhouse gases emissions. Collaborative NGOs that were heavily involved in this process expressed their satisfaction with the results of these negotiations because they were able 'to strengthen wording on GHG and peat lands and avoid … high carbon stock land'.[18] The reality, however, is that without tangible objectives and a clear deadline, producers can postpone any effort in the name of feasibility. Consequently, to implement these new GHG criteria, RSPO established yet another working group—the Emission Reduction Working Group.

Opponent NGOs' Strategies and Their Limitations

Opponent NGOs seek to protect tropical forests though a no deforestation commitment. These are powerful Western NGOs, among them Greenpeace and Friends of the Earth. They do not consider the RSPO as a credible means to produce sustainably. As an example, the Greenpeace report, *Certifying Destruction*, explains why companies need to go beyond the RSPO commitment (Greenpeace 2013). Opponent NGOs use the RSPO as a platform to expose bad practises in the supply chain. They are not RSPO members to avoid legitimising the organisation.

Opponent NGOs mobilise considerable resources to influence the RSPO to their advantage. First, they can mobilise (scientific) expertise, by investigating and establishing evidence against prominent growers who breach RSPO rules (e.g., Wilmar, Golden-Agri Resources). Second, with their close relationship to the media and the public, they can undertake aggressive campaigning that targets key RSPO members using a wide range of media tools (e.g., the video *Nestlé Killer*) or fact sheets such as 'How Unilever Suppliers Are Burning up Borneo'. Because they are not RSPO members, opponents cannot create RSPO rules or take strategic positions within it. For that reason, opponent NGOs benefit from the work of collaborative NGOs that can put forward decisions and hold strategic positions within the RSPO.

The strategic collaboration between the opponent and the collaborative NGOs began when Greenpeace made a public case against the Malaysian grower United Plantation that had breached the rules to implement the RSPO guidance document (Greenpeace 2008). In response to the case, the WWF put forward a resolution on new planting procedures at the 2008 GA. The decision forces growers to engage in public consultation for all new plantation permits, which are the state authorisations to plant oil palm on long-term governmental lease. Growers must publish the key information on their plantations' plan on the RSPO website for a duration of 30 days and need to include a map with the coordinates, a summary of the environmental impact assessment, and a summary of the development plan. With this resolution, opponent NGOs have had access to far more information on the RSPO growers. Accordingly, they could file many more cases against RSPO growers; as of April 2015, 50 cases had been brought to the RSPO (RSPO 2015c). The rising number of cases were new opportunities for the collaborative NGOs, especially Oxfam Novib and Sawit Watch. They could influence the RSPO to set up a formal grievance system open to members and nonmembers.

At first glance, collaborative and opponent NGO strategies generated a virtuous circle that strengthened the grievance system, facilitating the work of the opponent NGOs to identify new cases. Nevertheless, the system puts the burden on opponent NGOs, as they must sustain their claim through the slow procedure, often lasting for about two years for each case, because of the participatory and consensual approach within the RSPO. In addition, these NGOs can only focus on some specific cases, pooling their resources where it will have the greatest political impact. Indeed, growers are powerful actors with wide technical, financial and political resources.

Growers can turn low-income local communities against NGOs by compensating them for the land loss. They also can use comprehensive marketing strategies with a dedicated communications department (Wilmar 2011; GAR 2011). This can discredit an NGO-led international campaign. As a result of these constraints, opponent NGOs acknowledge: 'We are targeting the big one to make that big move. You know that we have been campaigning on XXX ... and it's because they are the biggest one, and then the biggest threat'.[19]

Opportunistic NGOs' Strategies and Their Limitations

The opportunistic NGOs are those RSPO members that seek to protect specific species and their habitat. Operating in Indonesia and Malaysia, they mainly focus on the Sumatran Orangutan (e.g., PanEco and SOS) and the Borneo Orangutan (e.g., Borneo Orangutan Society, Hutan and Orangutan Land Trust). Their financial and human resources are rather limited, but they benefit from a long-term scientific expertise with comprehensive knowledge about the field.

Initially, the opportunistic NGOs were rather unpredictable in the RSPO because they adopted opponent or collaborative strategies depending on the situation. Behaving as collaborative NGOs, they put forward several decisions during the RSPO GA to preserve specific territories, including cultivating palm oil on degraded land (PanEco in 2006), conserving the Tripa peat swamp forest (PanEco in 2008), conserving the Bukit Tigah Puluh ecosystem (SOS in 2009), and conserving nonprimary forest (SOS in 2010). In the GAs, opportunistic NGOs can be very hard on the RSPO system. PanEco's resolution on Tripa stated that 'we request the RSPO to adhere to the credibility of its role and drastically improve its efficiency by implementing an effective mechanism to control bad practices of the palm oil industry' (RSPO 2008: 39).

Remarkably, opportunistic NGOs also have put forward GA decisions of a much boarder scope than the SOS decision of 'Transparency in Plantations Concession Boundaries' adopted in 2013. This decision requests growers to provide the coordinates of each of its concessions, allowing NGOs to monitor what is happening on the ground in real time through remote sensing. They also actively participate on the various bodies that the RSPO has established to deal with biodiversity-related decisions (e.g., the BHCV WG and Compensation Task Force). Unlike the collaborative NGOs, however, they do not take strategic positions as cochairs but provide more scientific research-based evidence as participants. Behaving as opponent NGOs, so-called opportunistic NGOs also file complaint cases against RSPO members, produce fact-finding evidence and seek media exposure.

Changing their role to get the most impact, opportunistic NGOs are unpredictable within the RSPO. By institutionalising, the RSPO dramatically reduced the possibility for opportunistic NGOs to change roles. Growers successfully submitted the decision, 'Preserving the Integrity of the Standard Setting Process in RSPO', at the 2010 General Assembly. This decision forbids opportunistic NGOs from putting forward decisions that support their specific agenda. In addition, the RSPO created the codes of conduct, one for all their members and another for the BHCV WG members; these codes limit opportunistic strategies (RSPO 2015d, e). The BHCV WG Code of Conduct requests that the member's 'participation reinforces positive public image of the RSPO' (RSPO 2015e: 1), and it details a section about confidentiality for members in the working group. In other words, the Code of Conduct prevents opportunistic NGOs from adopting an opponent strategy or from collaborating with opponent NGOs, if the discussions as collaborative NGOs within the BHCV WG do not veer to their advantage.

As a result of the RSPO's institutionalisation, opportunistic NGOs have to choose between a collaborative and an opponent strategy, both limiting their impact on the RSPO. Involved with the RSPO for more than 10 years, most of these NGOs tend to adopt a collaborative strategy, sometimes reluctantly. For example, SOS continued its membership despite announcing their departure from the RSPO in November 2014.

Sceptic NGOs' Strategies and Their Limitations

Sceptic NGOs seek recognition of the communities' right to land and therefore conserve biodiversity and forests. Indeed, local communities have sustainably managed those biodiversity rich forest areas that are now threatened by oil palm plantations (Putri 2010). These sceptics are mostly operating in the producing countries with long-term social and environmental agendas (e.g., the WCS, the Leuser International Foundation, Sawit Watch, Whali, Forest Peoples Programme). Most of them are not RSPO members, expect for Forest Peoples Programme and Sawit Watch, which is a network of 40 social and environmental Indonesian NGOs (Jiwan 2013).

As the main sceptic NGO, the RSPO member, Sawit Watch, used three main avenues in the RSPO to reach its goals. First, it put forward GA decisions, such as 'The Need for a Task Force on Smallholders', at the GA2, cosigning them with international NGOs, producers and downstream firms. Second, it held the strategic position of representative of the social NGOs at the Executive Board to influence the implementation of these decisions. Third, it maintained strategic relationships at the grassroots level, facilitating the participation of the locals in the RSPO. Local participation took three forms: They brought violation cases of the RSPO standard against growers to the complaint's panel, they expressed their views at the RSPO annual meeting, and they organised mass protests during the annual meetings for those weak on RSPO standards for social issues (Parker 2013).

Despite the anecdotal achievements of getting the minority voice heard during the roundtable meetings, this strategy did not achieve its goals. The prevalent management discourse within the RSPO process discredited this land rights-based argument (Cheyns 2014). The RSPO's institutionalisation strengthened this structural problem over time by widening the discursive gap. By adopting this discourse, collaborative and opponent NGOs also marginalised the sceptic NGOs and their conservation agenda within the RSPO system.

It is, therefore, not a surprise that Sawit Watch gradually lost interest in the RSPO scheme. As a main consequence, it left the RSPO Biodiversity Group in 2012. Only Forest Peoples Programme has remained firmly active, focussing on the Free, Prior and Informed Concerned (FPIC) indicator of the guidance document (Cholchester and Chao 2014). If applied, the sceptic NGOs believe that the FPIC could be a powerful tool to get locals their rights to land recognised as on oil palm areas.

In conclusion, sceptics tend to avoid links with the RSPO because of their inability to influence it to truly include land rights issues. The RSPO tends to become secondary in their overall strategy. The sceptics distrust this initiative that diverts their limited resources. Working closely in the field, sceptics do not believe the RSPO can deal with the underlining sociopolitical–economic roots of forest destruction such as the lack of local land rights recognition, the high short-term economic return, the lack of law enforcement, and the promotion of the large-scale agricultural model. They remain highly sceptical in regard to the sudden conversion to the sustainability agenda by all the big producers. A representative of one of the NGOs summarises the general feeling: 'I have told you since the beginning. It's a waste of time [the RSPO]. It's pure brain washing. I haven't seen any of these oil palm people taken to court'.[20]

Concluding Remarks and Perspectives

The research discussed in this chapter demonstrates the importance of taking a political ecology approach and considering the diversity of NGOs' engagement with the RSPO. As strategic players with clear goals mobilising their limited resources, conservation NGOs use four broad strategies (or roles) to make conservation gains: Collaborative approaches seek to change the system from within by providing scientific research-based information, holding strategic positions and creating rules; opponent NGOs remain outside the RSPO while using it as a platform for public campaigns; opportunistic strategies focus on conserving geographical areas, adopting either collaborative or opponent strategies to reach their goals; finally, those adopting the role of the 'sceptic' take a rights-based approach to conservation by working with local communities to stop oil palm plantation expansion.

Within these strategies, conservation NGOs play a vital role in strengthening biodiversity conservation within the RSPO. This includes the adoption of many GA decisions; the creation of a comprehensive complaints system; and the BHCV WG, a Compensation Task Force and a Working Group on Greenhouse Gases. All these rules and entities help to strengthen the RSPO standard and the provisions for its implementation.

Nevertheless, the RSPO system has largely failed to support NGOs in reaching their initial conservation goals. The ongoing mobilisation of NGO resources and responses from other players, especially growers and downstream actors (e.g., processors, manufacturers, retailers), has co-constructed and institutionalised the RSPO in such a way that it constrains NGOs. This limits NGOs' options, preventing them from reaching their initial goals. As such, NGOs have obtained only some very minor conservation outcomes. Three additional reasons for this were found. The initial reason is that institutionalisation forces each individual NGO to keep to a specific strategy while their resources are decreasing whereas the RSPO is structuring. Then, the institutionalisation impedes collaboration between NGOs using various strategies, while this collaboration is key to getting leverage. Finally, sceptic NGOs are structurally excluded because of their incompatibility with the prevalent capitalistic discourse within the RSPO, while their participation is fundamental both in terms of long-term conservation gains and social justice.

Conservation NGOs are directly responsible for this situation. Indeed, the RSPO system tends to promote a collaborative strategy that uses scientific research-based managerial discourses with a focus on technologies such as remote sensing. Collaborative NGOs fully engage in providing scientifically based information on biodiversity loss, deforestation or climate change. Still, those NGOs are never able to provide all the needed information. There are always some measurement uncertainties at the local level. In addition, the fundamental consensus-building rule itself promotes the market logic (Büscher 2008). Indeed, each agreement in a working group is incomplete with ongoing adjustments of the means and goals (McCarthy 2012).

As such, the working groups that conservation NGOs have succeeded in establishing evolved into long-term institutions. They digest NGO criticism through a science-based management process (Boltanski and Chiapello 1999). As a result, the RSPO still allows biodiversity loss, deforestation and

GHG emissions. Having contributed to reinforcing the RSPO system for more than 10 years, collaborative NGOs seem to have no other option than to unconditionally promote the RSPO. For instance, in 2013, an NGO advocated in favour of the RSPO guidance document endorsed at the 2013 GA despite its flaws: 'On balance the new P&Cs [Principles & Criteria] are not the end point but a step on the way to transforming the industry and as such, I think we should support the revisions [of the guidance document] and encourage others to do so'.[21]

Opponent NGOs initially cooperate with collaborative NGOs to expose bad practises of RSPO members and to reinforce the RSPO complaints system. Nonetheless, opponent NGOs can focus only on specific cases against the largest RSPO growers by pooling their limited resources where it will have the greatest political impact. As a result, opponent NGOs can only play the role of external supervisor, a role normally given to the state, without having the financial and coercive power of the state. Opportunistic NGOs initially use collaborative or opponent strategies within the RSPO to protect specific geographical areas. With the RSPO's institutionalisation, however, they are forced to choose between the two strategies because new RSPO rules force all members to promote it. Reluctantly, the NGOs tend to select the collaborative strategy.

Finally, NGOs adopting the role of the sceptic seek recognition of local land rights but their requests are marginalised and not taken into consideration within the RSPO system. They tend to leave the RSPO, even though land rights, and more broadly local issues, are a fundamental matter of concern for biodiversity conservation. The NGOs are also unable to get leverage on the RSPO from developing collaborative strategies between different NGO categories. This is because the RSPO has adopted two formal codes of conduct, one for its members and one for the BHCV WG, that force all RSPO members to adopt only collaborative strategies.

In summary, by analysing NGO strategies relative to the RSPO and the main actors' agendas, and how NGOs participate or oppose based on their own motivations, this research expands our understanding of why conservation NGOs obtain meagre conservation outcomes. In the case of the RSPO, it results from the growing gap between the intent (the initial goal of NGOs) and the allowed actions in an always more constrained environment. What appear as 'alliances' (Igoe et al. 2010a: 486) between growing markets and conservation may be experienced by the NGOs as their only possible choice, or at least the most feasible one.

Yet, what can the NGOs do to get better conservation outcomes? One option would be for NGOs to clearly focus on their initial conservation objective by mobilising additional resources in a constrained system. Another option would be to collaborate on a collective conservation or human rights objective with other NGOs using diverse strategies in order to get some leverage (Ruysschaert 2013). At the field level, sceptic NGOs could support local communities to regain recognition of their land rights. Along the supply chain, collaborative NGOs could focus on transparency and traceability. The World Resources Institute with its Global Forest Watch and ZSL with its Sustainable Palm Oil Platform (ZSL 2015) are already actively participating in this. At the consumer level, opponent NGOs could expose the bad practises of main players (e.g., producers, traders, financial institutions) by using information that has been made available by collaborative NGOs. Finally, opportunistic NGOs could benefit from the work of the other NGOs and apply it to specific territories.

To test whether the NGOs would benefit more from focussing on their specific objective or from collaborative strategies to get an impact, researchers should conduct a more detailed institutional ethnographic study of a selection of NGOs (Larsen 2016), which is beyond the scope of this chapter. An ethnographic analysis of the NGOs involved in the Palm Oil Innovative Group (POIG) may be a promising avenue (POIG 2013). The POIG brings together several proactive RSPO growers and NGOs from the four engagement strategies: the WWF (collaborative), Greenpeace (opponent), Forest Peoples Programmes (sceptics) and SOS (opportunistic). The POIG members commit to respecting both human rights and to 'no deforestation', which includes the preservation of all peat lands and forests above a certain carbon stock threshold.

Acknowledgments This chapter draws heavily on Denis Ruysschaert's Ph.D. thesis on the role of conservation NGOs in biodiversity loss (Ruysschaert 2013). Thank you to the three anonymous reviewers and the editor for the helpful comments. Thank you to Vi Tran, Stephen Caudwell and Mira Gratier for their English editing. Thank you to Satu Ojaluoma for her moral support.

Notes

1. Aquaculture Stewardship Council, Better Cotton Initiative—BCI, Roundtable on Sustainable Biofuels—RSB, Better Sugar Cane Initiative—Bonsucro, Roundtable on Responsible Soy—RTRS.
2. (1) Commitment to transparency, (2) compliance with applicable laws and regulations, (3) commitment to long-term economic and financial viability, (4) use of appropriate best practises by growers and millers, (5) environmental responsibility and conservation of natural resources and

biodiversity, (6) responsible consideration of employees and of individuals and communities affected by growers and mills, (7) responsible development of new plantings, and (8) commitment to continuous improvement in key areas of activity.

3. Environmental NGO interview, 20 April 2012.
4. Retailer funding member RSPO, 10 December 2012.
5. Interview, 28 November 2011.
6. SOS interview, 18 March 2012.
7. Grower interview, 13 February 2012.
8. Sime Darby, the largest oil palm grower in the world, bought New Britain Oil Palm in March 2016.
9. Grower interview, 13 February 2012.
10. Grower interview, 13 February 2012.
11. Interview, 18 November 2011.
12. Votes were 95 for, 4 against, 22 abstentions.
13. Three growers from Indonesia (PT Smart, Musim Mas, Gapki) and three from Malaysia (IOI, Kulim Bernhard, Sime Darby, Wilmar).
14. Indonesian growers (PT Indonesia Plantations, Asian Agri Group, GAPKI, PT Smart) and Malaysian growers (United Plantations, Sime Darby).
15. Conservation International, Wetlands International, WWF International, WWF Malaysia, and WWF Indonesia.
16. Palanka Raya Indonesia University, Parum Agricultural Soil Survey, Delft Hydraulics, Leicester University.
17. Research Institute interview, 10 November 2011.
18. Interview, 28 November 2011.
19. Interview, 23 November 2011.
20. Interview, 19 November 2011.
21. E-mail to all NGOs regarding RSPO vote on the new P&C, 9 April 2013.

References

Agus, F., I.E. Henson, B.H. Sahardjo, N. Harris, M. van Noordwijk, and T.J. Killeen. 2013. Review of Emission factors for Assessment of CO^2 Emission from Land Use Change to Oil Palm in Southeast Asia. In *Reports from the Technical Panels of the Second RSPO GHG Working Group*, ed. T.J. Killeen and J. Goon, 7–31. Kuala Lumpur: RSPO.

Barker, M. 2009. When Environmentalists Legitimize Plunder. *Swans Commentary* 15. http://www.swans.com/library/art15/barker12.html. Accessed 5 Oct 2015.

Benedick, S., J. Hill, N. Mustaffa, V. Chey, M. Maryati, J. Searle, M. Schilthuizen, and K. Hamer. 2006. Impacts of Rainforest Fragmentation on Butterflies in Northern Borneo: Species Richness, Turnover and the Value of Small Fragments. *Journal of Applied Ecology* 43 (5): 967–977.

Benjaminsen, T., and H. Svarstad. 2010. Questioning Conservation Practice—and Its Response: The Establishment of Namaqua National Park. *Current Conservation* 3 (3): 8–13.

Boltanski, L., and E. Chiapello. 1999. *Le nouvel esprit du capitalisme*. Paris: Gallimard.

Brockington, D. 2011. A Brief Guide to (Conservation) NGOs. *Current Conservation* 5 (1): 30–31.

Brockington, D., and R. Duffy. 2010. Capitalism and Conservation. The Production and Reproduction of Biodiversity Conservation. *Antipode* 42 (3): 469–484.

Büscher, B. 2008. Conservation, Neoliberalism, and Social Science: A Critical Reflection on the SCB 2007. Annual Meeting in South Africa. *Conservation Biology* 22 (2): 229–231.

Büscher, B., W. Dressler, and R. Fletcher, eds. 2014. *Nature Inc.: Environmental Conservation in the Neoliberal Age*. Tucson: University of Arizona Press.

Chaplin, M. 2004. A Challenge to Conservationists. *World Watch Magazine* 17 (6): 17–31.

Cheyns, E. 2012. (Dé) politisation des standards dans les dispositifs de normalisation multiparties prenantes. Les cas du soja et de l'huile de palme. In *Normaliser au nom du développement durable*, ed. P. Alphandréry, M. Djama, and A. Fortier, 101–118. Versailles: Quae.

———. 2014. Making "Minority Voices" Head in Transnational Round Tables: The Role of NGOs in Reintroducing Justice and Attachments. *Agriculture and Human Values* 31 (3): 439–453.

Cholchester, M., and S. Chao. 2014. *Revising the RSPO Guide on Free, Prior and Informed Consent. Conclusions and Questions. RT12*. Moreton-in-Marsh: Forest Peoples Programme.

Colchester, M., and S. Chao, eds. 2011. *Oil Palm Expansion in South East Asia. Trends and Implications for Local Communities and Indigenous Peoples*. Moreton-in-Marsh: Forest Peoples Programme and Sawit Watch.

Crozier, M., and E. Frieberg. 1977. *L'Acteur et le Système*. Paris: Seuil.

DGEC. 2014. *2013–2015 Tree Crop Estate Statistics of Indonesia. Palm Oil*. Jakarta: Directorate General of Estate Crops.

Edwards, D., J. Hodgson, K. Hamer, S. Mitchell, A. Ahmad, S. Cornell, and D. Wilcove. 2010. Wildlife-Friendly Oil Palm Plantations Fail to Protect Biodiversity Effectively. *Conservation Letters* 3 (4): 236–242.

Escobar, M., and E. Cheyns. 2012. NGOs Acting Along Global Supply Chains: Between Market and Politics. An Assessment of Campaign and Regulatory Devices in the Palm Oil and Soy Sectors. In *Workshop: Responsible Supply Chains and Networks: Challenges for Governance and Sustainability*, 22–24 November. Stockholm: Södertörn University.

Fitzherbert, E., M. Struebig, A. Morel, F. Danielsen, A. Brühl, P. Donald, and B. Phalan. 2008. How Will Oil Palm Expansion Affect Biodiversity? *Trends Ecology and Evolution* 23 (10): 538–545.

Fouilleux, E. 2013. Normes Transnationales de Développement Durable. *Gouvernement & Action Publique* 2 (1): 93–119.

Fouilleux, E., and F. Goulet. 2012. Firmes et Développement Durable: le Nouvel Esprit du Productivisme. *Etudes Rurales* 190: 131–146.

Friedberg, E. 1991. *Le pouvoir et la règle. Dynamiques de l'action organisée*. Paris: Seuil.

GAR. 2011. *Forest Conservation Policy*. Kuala Lumpur: Golden-Agri Resources.

Gautier, D., and T. Benjaminsen. 2012. Introduction à la Political Eology. In *Environnement, discours et pouvoir. L'approche Political Ecology*, ed. D. Gautier and T. Benjaminsen, 5–20. Versailles: Quae.

Greenpeace. 2008. *United Plantations Certified Despite Gross Violations of RSPO Standards*. Amsterdam: Greenpeace International. http://www.greenpeace.org/international/Global/international/planet-2/report/2008/11/united-plantations-certified-d.pdf. Accessed 10 Mar 2015.

———. 2013. *Certifying Destruction. Why Consumer Companies Need to Go Beyond the RSPO to Stop Forest Destruction*. Amsterdam: Greenpeace International. http://www.greenpeace.de/files/publications/rspo-certifying-destruction.pdf. Accessed 10 Mar 2015.

High Commerical Value (HCV). 2007. *Expert Workshop on Methods for Landscape HCV*, 12–13 February. Bogor: WWF.

Hooijer, A., M. Silvius, H. Wösten, and S. Page. 2006. *PEAT-CO^2: Assessment of CO^2 Emissions from Drained Peatlands in SE Asia*. Delft, the Netherlands: Delft Hydraulics. http://www.stockandland.com.au/news/agriculture/agribusiness/general-news/adm-ups-wilmar-stake/2719170.aspx 09. Accessed 29 Apr 2015.

Igoe, J. 2005. Global Indigenism and Spaceship Earth: Convergence, Space, and Reentry Friction. *Globalizations* 3 (2): 377–390.

Igoe, J., and D. Brockington. 2007. Neoliberal Conservation: A Brief Introduction. *Conservation and Society* 5 (4): 432–449.

Igoe, J., K. Neves, and D. Brockington. 2010a. A Spectacular Eco-tour Around the Historic Bloc: Theorising the Convergence of Biodiversity Conservation and Capitalist Expansion. *Antipode* 42 (3): 486–512.

Igoe, J., S. Sullivan, and D. Brockington. 2010b. Problematising Neoliberal Biodiversity Conservation: Displaced and Disobedient Knowledge. *Current Conservation* 3 (3): 4–7.

IUCN. 2015. IUCN Red List of Threatened Species. www.iucnredlist.org. Accessed 25 June 2015.

Jiwan, N. 2013. The Political Ecology of the Indonesia Palm Oil Industry, a Critical Analysis. In *The Palm Oil Controversy in South East Asia. A Transnational Perspective*, ed. O. Pye and J. Bhattacharya, 48–75. Singapore: Institute of Southeast Asian Studies.

Killeen, T.J., and J. Goon, eds. 2013. *Reports from the Technical Panels of the Second RSPO GHG Working Group*. Kuala Lumpur: RSPO.

Larsen, P. 2016. The Good, the Ugly and the Dirty Harry's of Conservation: Rethinking the Anthropology of Conservation NGOs. *Conservation and Society* 14 (1): 21–33.

Margono, B., P. Potapov, S. Turubanova, F. Stolle, and M. Hansen. 2014. Primary Forest Cover Loss in Indonesia over 2000–2012. *Nature Climate Change* 4: 730–735.

McCarthy, J. 2002. First World Political Ecology: Lessons from the Wise Use Movement. *Environment and Planning A* 34 (7): 1281–1302.

———. 2012. Certifying in Contested Spaces: Private Regulation in Indonesian Forestry and Palm Oil. *Third World Quarterly* 33 (10): 1871–1888.

Miettinen, J., A. Hooijer, S. Chenghua, D. Tollenaar, R. Vernimmen, S. Chin Liew, C. Mallins, and S. Page. 2012. Extent of Industrial Plantations on Southeast Asian Peatlands in 2010 with Analysis of Historical Expansion and Future Projections. *GCB Bioenergy* 4: 908–918.

Oosterveer, P. 2014. Promoting Sustainable Palm Oil: Viewed from a Global Networks and Flows Perspective. *Journal of Cleaner Production* 107: 1–8.

Parker, D. 2013. Palm Oil Workers Stage Massive Protest at Sustainability Meeting in Indonesia. *Mongabay*, November 19. http://news.mongabay.com/2013/1119-dparker-palm-oil-worker-protest.html. Accessed 13 June 2014.

Peluso, N. 2012. Situer les Political Écologies: l'Exemple du Caoutchouc. In *Environnement, discours et pouvoir. L'approche Political Ecology*, ed. D. Gautier and T. Benjaminsen, 37–65. Versailles: Quae.

Persey, S., R. Nussbaum, M. Hatchwell, S. Christie, and H. Crowley. 2011. *Towards Sustainable Palm Oil: A Framework for Action*. Cambridge: ZSL/Proforest/WCS.

POIG. 2013. *Palm Oil Innovations Group Charter*. Gland, Switzerland: WWF. http://www.greenpeace.org/international/Global/international/photos/forests/2013/Indonesia%20Forests/POIG%20Charter%2013%20November%202013.pdf. Accessed 14 June 2015.

Ponte, S., J. Vestergaad, and P. Gibbon, eds. 2011. *Governing Through Standards: Origins, Drivers and Limits*. London: Palgraves.

Putri, N. 2010. Building a State, Dispossessing the Nation: Sovereignty and Land Dispossession in Indonesia. Ph.D. thesis. Graduate Institute, Geneva, Switzerland.

Pye, O. 2013. The Political Ecology of the Indonesia Palm Oil Industry, a Critical Analysis. In *The Palm Oil Controversy in South East Asia. A Transnational Perspective*, ed. O. Pye and J. Bhattacharya, 179–198. Singapore: ISEAS Publishing.

Pye, O., and J. Bhattacharya. 2013. *The Palm Oil Controversy in Southeast Asia. A Transnational Perspective*. Singapore: ISEAS Publishing.

RSPO. 2002. *Minutes of the Preparatory Meeting*, 20 September. London: RSPO.

———. 2004. *Statutes—Roundtable on Sustainable Palm Oil*. Kuala Lumpur: RSPO.

———. 2007. *RSPO Certification Systems. Final Document Approved by RSPO Executive Board*, 26 June. Kuala Lumpur: RSPO. http://www.rspo.org/sites/default/files/RSPOcertification-systems.pdf. Accessed 20 Mar 2014.

———. 2008. *Minutes of the 5th General Assembly of the Roundtable on Sustainable Palm Oil*. Kuala Lumpur: RSPO. http://www.rspo.org/files/resource_centre/MinutesGA5.pdf. Accessed 20 Mar 2014.

———. 2010. *Roundtable on Sustainable Palm Oil 7th General Assembly*. Kuala Lumpur: RSPO. http://www.rspo.org/file/GA7%20Minutes.pdf. Accessed 20 Mar 2014.

———. 2012. *Roundtable on Sustainable Palm Oil 8th General Assembly. New Vision and Mission Statements for the RSPO. Resolution 6b*. Kuala Lumpur: RSPO. http://www.rspo.org/resources/supplementary-materials/minutes-reports-of-rspo-ga-ega. Accessed 21 June 2016.

———. 2013. *Principles and Criteria for the Production of Sustainable Palm Oil*. KualaLumpur:RSPO.www.rspo.org/publications/download/224fa0187afb4b7. Accessed 21 June 2016.

———. 2014. *Task Force Compensation Recommendations*. Kuala Lumpur: RSPO. http://www.rspo.org/file/2_RSPORemediationandCompensationProcedures_May2014.pdf. Accessed 21 June 2016.

———. 2015a. *Members*. Kuala Lumpur: RSPO. http://www.rspo.org/members/all. Accessed 20 May 2015.

———. 2015b. Impact. Latest update on 17 June 2015. http://www.rspo.org/about/impacts. Accessed 5 July 2015.

———. 2015c. *Complaints*. Kuala Lumpur: RSPO. http://www.rspo.org/members/complaints. Accessed 29 Apr 2015.

———. 2015d. *Code of Conduct for Members of the RSPO*. Kuala Lumpur: RSPO. http://www.rspo.org/files/resource_centre/CoC.pdf. Accessed 21 June 2016.

———. 2015e. *Code of Conduct Policy Statement—RSPO BHCV Working Group*. Kuala Lumpur: RSPO. http://www.rspo.org/about/who-we-are/working-groups/biodiversity-high-conservation-values. Accessed 21 June 2016.

———. 2015f. *Biodiversity and High Conservation Value Working Group*. Kuala Lumpur: RSPO. http://www.rspo.org/about/who-we-are/working-groups/biodiversity-high-conservation-values. Accessed 10 May 2015.

Ruysschaert, D. 2013. Le rôle des organisations de conservation dans la construction et la mise en oeuvre de l'agenda international de conservation d'espèces emblématiques: le cas des orangs-outans de Sumatra. Ph.D. thesis. University Jean Jaurès, Toulouse, France.

———. 2016. 1. The Impact of Global Palm Oil Certification on Transnational Governance, Human Livelihoods and Biodiversity Conservation. *IUCN Policy Matters* 21: 45–58.

Ruysschaert, D., and D. Salles. 2014. Towards Global Voluntary Standards: Questioning the Effectiveness in Attaining Conservation Goals. The Case of the Roundtable on Sustainable Palm Oil (RSPO). *Ecological Economics* 107: 438–446.

Ruysschaert, D., A. Darsoyo, R. Zen, G. Gea, and I. Singleton. 2011. *Developing Palm-Oil Production on Degraded Land*. Medan: PanEco Foundation, YEL and World Agroforestry Centre.

Sachedina, H., J. Igoe, and D. Brockington. 2010. The Spectacular Growth of a Conservation NGO and the Paradoxes of Neoliberal Conservation. *Current Conservation* 3 (3): 24–27.

Struebig, M., T. Kingston, A. Zubaid, A. Mohd-Adnan, and S. Rossiter. 2008. Conservation Value of Forest Fragments to Palaeotropical Bats. *Biological Conservation* 141 (8): 2112–2126.

Struebig, M., T. Kingston, E. Petit, S. Le Comber, A. Zubaid, A. Mohd-Adnan, and S. Rossiter. 2011. Parallel Declines in Species and Genetic Diversity in Tropical Forest Fragments. *Ecology Letters* 14 (6): 582–590.

Tumusiime, D., and H. Svarstad. 2011. A Local Counter-Narrative on the Conservation of Mountain Gorillas. *Forum for Development Studies* 38 (3): 239–265.

USDA. 2014. United States Department of Agriculture PSD Database. *Agricultural Production, Supply, and Distribution*. Washington, DC: USDA. http://www.indexmundi.com/agriculture/. Accessed 25 June 2014.

Wetlands International. Amended Resolution 19 November 2008. 5th RSPO General Assembly. Wageningen, the Netherlands: Wetlands International.

Wich, S.A., J. Riswan, J. Refisch Jenson, and C. Nellemann, eds. 2011. *Orangutans and the Economics of Sustainable Forest Management in Sumatra*. Trykkeri: GRASP/ YEL/ ICRAF/ GRID-Arendal.

Wilmar International. 2011. *Sustainability Report 2011: Staying the Course Through Challenging Times*. Singapore: Wilmar International.

WWF. 2012. *Market Transformation Initiative Strategy*. Gland, Switzerland: WWF International. http://awsassets.panda.org/downloads/market_transformation_initiative_strategy_1.pdf. Accessed 29 Apr 2015.

———. 2013. *WWF Assessment of RSPO Member Palm Oil Producers 2013*. Gland: WWF International. http://wwf.panda.org/what_we_do/footprint/agriculture/palm_oil/solutions/responsible_purchasing/wwf_assessment_of_rspo_member_palm_oil_producers_2013/. Accessed 29 Apr 2015.

WWF, ZSL and GFN. 2014. *WWF Living Planet Report 2014. Species and Paces, People and Places*. Gland: WWF, Zoological Society of London and Global Footprint Network. http://wwf.panda.org/about_our_earth/all_publications/living_planet_report/. Accessed 5 July 2015.

ZSL. 2015. *Sustainable Palm Oil Platform*. Cambridge: Zoological Society of London. http://www.sustainablepalmoil.org/. Accessed 29 Apr 2015.

CHAPTER 6

Investigating the Consistency of a Pro-market Perspective Amongst Conservationists

Libby Blanchard, Chris G. Sandbrook, Janet A. Fisher, and Bhaskar Vira

We thank the staff of various conservation organisations in Cambridge, UK, for participating in the Q survey interviews for this study, and the Gates Cambridge Trust for supporting the first author's research. We also thank the three anonymous reviewers of this chapter for their insightful comments, which have considerably sharpened our argument.

L. Blanchard (✉) • B. Vira
Department of Geography, University of Cambridge, Cambridge, UK

C.G. Sandbrook
UNEP World Conservation Monitoring Centre Cambridge, Cambridge, UK

Department of Geography, University of Cambridge, Cambridge, UK

J.A. Fisher
School of Geosciences, University of Edinburgh, Edinburgh, UK

P.B. Larsen, D. Brockington (eds.), *The Anthropology of Conservation NGOs*, Palgrave Studies in Anthropology of Sustainability, DOI 10.1007/978-3-319-60579-1_6

Introduction

Neoliberal Conservation and Its Critics

In the last two decades, the field of biodiversity conservation has increasingly been characterised by the use of market logic (MacDonald 2010b; Büscher et al. 2012; Pirard 2012). Even though so-called 'market-based instruments' (MBIs), such as ecotourism, taxation and subsidies, have existed in conservation practise for quite some time, new instruments, such as payments for ecosystem services (PES), biodiversity derivatives and offsets and mitigation banking, have recently become more widespread (Landell-Mills and Porras 2002; Pattanayak et al. 2010; Arsel and Büscher 2012). Market-based instruments can be defined as mechanisms that attribute a price to nature (Pirard 2012), although their links to true markets are sometimes questionable (Pirard 2012; Muradian et al. 2013). Increasingly, MBIs are used in conservation because there is an expectation that they will deliver: (1) efficiency through the use of the market to internalise externalities (Brockington and Duffy 2011), (2) an economic rationale for conservation that decision makers understand (Daily 1997; Pearce and Barbier 2000), (3) new funding sources (Ferraro 2001; Balmford and Whitten 2003; Wunder 2007), and (4) potential "win–wins" by addressing both biodiversity conservation and poverty alleviation (Pagiola et al. 2005).

Market-based conservation practise has resulted in new and altered relationships between conservation organisations, the public and private sectors and local people (Sandbrook et al. 2013b). Mainstream environmental policy initiatives, including the study by 'The Economics of Ecosystems and Biodiversity' (TEEB) (TEEB 2010), the UN Environment Programme's 'Towards a Green Economy' report (UNEP 2011), the Convention on Biological Diversity's 'Strategic Plan for Biodiversity 2011–2020' paper (CBD 2010), the EU Biodiversity Strategy to 2020 report (European Commission 2011), and the UK's Natural Capital Committee document (www.naturalcapitalcommittee.org); these all promote the use of market-based instruments to conserve and manage nature.

Many biodiversity conservation organisations have increasingly promoted the use of markets and have formed partnerships with private sector actors (Igoe and Brockington 2007; MacDonald 2010b). These practises represent a departure from conservation norms that were prevalent up until two decades ago, when mainstream conservation organisations were more likely to embrace "values, approaches and missions that were deeply incompatible" with private sector interests (MacDonald 2010b: 515) and to espouse alternative perspectives challenging neoliberal capitalism (Harvey 2003).

This neoliberal turn in conservation and the presumed win–win claims advanced by such perspectives have been the subject of criticism from within the social science community (Igoe and Brockington 2007; Arsel and Büscher 2012). Neoliberal conservation is defined as an "amalgamation of ideology and techniques informed by the premise that natures can only be 'saved' through their submission to capital and its subsequent revaluation in capitalist terms" (Büscher et al. 2012: 4). Such scholars see this trend as part of the wider political economic process of neoliberalisation (McCarthy and Prudham 2004; Igoe and Brockington 2007; Castree 2008)—the expansion of markets into increasingly broad areas of society over the last few decades (Sandel 2012). Given that neoliberal conservation "privileges as a solution the very structures and processes of neoliberal capitalism that produce the socio-ecological damages it seeks to redress" (Büscher et al. 2012: 14), critical social scientists question whether capitalist market mechanisms can resolve environmental problems (Igoe and Brockington 2007; Storm 2009).

Specifically, critical social scientists interrogate the process by which neoliberal conservation commodifies nature and question the outcomes of this process, which some have dubbed "Nature™ Inc." (Arsel and Büscher 2012). Castree (2008) argues that neoliberalism does not so much entail deregulation as require reregulation to reduce complex ecosystems into tradable commodities. This commodification process creates new types of value (Arsel and Büscher 2012) and facilitates new avenues for capital accumulation (Schurman and Kelso 2003; Harvey 2006). In addition to the moral critique of whether nature should be reduced to tradable commodities, scholars question the outcomes of this commodification process, which can result in privatisation (Heynen and Robbins 2005), social and economic displacement or exclusion (Brockington 2002), and the potential for elite-capture of newly available benefits (Igoe and Brockington 2007).

Further, McAfee (2012) argues that even though much attention has been focussed on the technical and institutional obstacles to implementing market-based approaches to conservation, this has detracted from the ways in which winners and losers are actively produced through the exchange process of markets. Indeed, the act of trading itself has significant redistribution consequences (Vira 2002). Thus, scholars argue that instead of creating synergistic growth and win–wins, the neoliberalisation of nature could result in winners and losers and potentially a reduction in biodiversity conservation effectiveness (Vira 2015).

The Circulation of Ideas Within Epistemic Communities

The critiques advanced by social scientists regarding neoliberal conservation seem to have had little impact on conservation professionals or decision makers. Büscher et al. (2012) observe that neoliberal solutions among conservationists seem to be the result of a consensus and seem not to be subjected to internal dissent. They propose that this lack of dissent is because neoliberal conservation functions as an ideology that has become socially (and ecologically) embedded through the generation of hegemonic governance structures and practises (Büscher et al. 2012). They write, "as an ideology it needs to be believed in; its central tenets should not be questioned" (Büscher et al. 2012: 15). Thus, they propose that critical perspectives, when voiced among conservation professionals, are not just ignored but also are actively suppressed or 'muted' in various ways, including through "the disciplining force of denial or disregard" (Büscher et al. 2012: 21).

Peck (2011) calls such a phenomenon, where various forms of evidence are ignored and policies reflect the strategies of those with certain interests, the result of 'fast policy'. He notes that in such contexts, what appears to be pragmatic policy learning is in fact policy replication operating within specific ideological parameters, with a 'push' toward particular solutions among policy entrepreneurs and decision makers (Peck and Theodore 2010b). These authors write that "there can be no doubt that canalized forms of transnational policy learning ... embedded within tight ideological parameters—are playing heightened roles in animating" various policy landscapes (Peck and Theodore 2010b: 206).

The trend toward market-based practise in biodiversity conservation could be the result of shifting perspectives among conservationists as a 'community of practice', led by a similarly tight set of key decision makers. A community of practise (Lave and Wenger 1991; Wenger 2000) is defined as a "collection of people who engage on an ongoing basis in some common endeavor" (Eckert 2006: 683) and who share practises reflecting their collective learning (Wenger 2000; Eckert 2006). As a result, they develop shared ways of addressing recurring problems. Nonetheless, how do collective strategies within a community of practise shift and evolve? Who drives and decides what types of knowledge and tools are embraced or rejected within a community? Wenger (2000) and Peck and Theodore (2010b) refer to this process as an 'alignment' that coordinates perspectives, interpretations and action. Indeed, Peck and Theodore argue that such alignment comes from dense expert networks and epistemic communities of practise that "serve and sustain" policies (Peck and Theodore 2010b: 207).

Haas corroborates this link between alignment and epistemic communities, which he defines as "networks of knowledge-based experts" (Haas 1992: 2) who play key roles in articulating and framing problems and potential solutions within their particular communities. According to Haas (1992), such communities defer to these 'knowledge elites' who shape strategies and thinking within the community, resulting in the community of practise learning new patterns of reasoning and behaviour.

The propagation of ideas within epistemic communities usually involves close personal connections, shared ideologies and repeated interaction in a variety of settings. As such, international conference meetings and the geographic co-location of professionals are both examples of settings where such propagation and alignment of ideas could occur. Indeed, MacDonald (2010a) argues that high-profile meetings within the biodiversity conservation community provide important 'fields' for social and cultural reproduction; in other words, "moments when conservationists come together as a tangible (as opposed to imaginary) global community to reaffirm their values and beliefs" (Büscher et al. 2012: 18). This is also the case with geographic co-location.

Examined through the lens of this scholarship, one explanation for the increasing dominance of market-based interventions in conservation could be the emergence of an epistemic community, actively promoting the adoption of such approaches during the last two decades. Holmes (2011) suggests that such a community, a 'transnational conservation elite' that shares, replicates, amplifies and promotes ideas about market-based solutions to the conservation community, is emergent; however, he does not provide empirical evidence indicating the existence of such a transnational conservation elite.

Methods

This chapter uses Q methodology to empirically investigate perspectives on the role of markets in conservation among members of two distinct, but potentially overlapping, conservation communities. It does so by comparing the results of an earlier Q study conducted with delegates at the December 2011 International Congress for Conservation Biology (ICCB) in Auckland, New Zealand—hereafter referred to as 'the ICCB study' (Sandbrook et al. 2013b)—with new data collected using the same Q methodology instrument with staff members of conservation organisations based in and around Cambridge, UK—hereafter referred to as 'the Cambridge study'.

For this comparative study, we conducted semistructured interviews using Q methodology. The Q methodology increasingly has been used in the field of social science research on conservation (Sandbrook et al. 2013a) to study conservationists' subjective values and perspectives (Mazur and Asah 2013; Rastogi et al. 2013; Sandbrook et al. 2013b), as well as being used more and more in environmental research (e.g., Robbins 2000, 2006; López-i-Gelats et al. 2009; Brannstrom 2011). We employed the same Q methodology survey as the Sandbrook et al. (2013b) ICCB study, with 17 professionals working for conservation organisations located in and around Cambridge, UK, in February and March of 2013. We then conducted a direct comparison of the perspectives represented in both the ICCB and Cambridge datasets.

The city of Cambridge has one of the largest clusters of biodiversity conservation organisations and researchers in the world[1]; many are linked through the Cambridge Conservation Forum (CCF),[2] the Cambridge Conservation Initiative (CCI),[3] and the University of Cambridge Conservation Research Institute (UCCRI).[4] A number of staff of CCI member organisations are now housed together in a new campus building that serves as the hub for CCI (Stokstad 2016). This research included, but was not limited to, conservation professionals who were employees of conservation organisation partners of the Cambridge Conservation Initiative and affiliated with the University of Cambridge Conservation Research Institute.

The organisations with employees that took part in the interviews represent the various types, sizes and focusses of conservation organisations that exist in Cambridge. The Cambridge conservation community is made up of global intergovernmental organisations (IGOs) such as the United Nations Environment Programme World Conservation Monitoring Centre, the International Union for the Conservation of Nature, and the International Whaling Commission; internationally focussed, large biodiversity conservation organisations and networks such as Flora and Fauna International and Birdlife International; and much smaller conservation NGOs with a more limited remit and geographical focus.

The larger conservation organisations in Cambridge have more than 200 employees at their organisational headquarters, with partners and networks across more than 100–150 countries; their annual operational budgets are on the order of USD40–50 million. Those at the other end may have less than five employees and correspondingly smaller budgets. Many employees have Masters degrees (or equivalent) or PhDs. The Cambridge organisations represent what might be thought of as 'mainstream' conservation practise, and do not include any organisations associated with radical opposition to market-based conservation. Many individuals from the Cambridge conservation community (typically those who are more senior members of larger conservation organisations) interact with other members of the global conservation community through participation in conferences, knowledge platforms and global convention meetings, as well as through employment mobility.

Nevertheless, the Cambridge conservation community also includes networks and initiatives, developed over approximately the last 15 years, that are designed to facilitate frequent interactions between conservation professionals, conservation science researchers and critical social scientists. Therefore, Cambridge-based conservation professionals and scientists might be more exposed to critical thinking on markets and conservation than is the norm in other contexts. The research for this chapter, therefore, sheds some light on whether this slightly unusual cluster of individuals and organisations reproduces or departs from the patterns observed at a gathering of conservation professionals at ICCB 2011.

Q Methodology

To undertake this research, we used Q methodology, a method used to quantitatively study respondents' subjectivity and first-person viewpoints (Watts and Stenner 2012). This methodology statistically measures the extent to which opinions and viewpoints are shared between respondents (Watts and Stenner 2012). It combines the qualitative study of perceptions with statistical analysis, revealing key qualitative viewpoints and allowing them to be understood at a high level of quantitative detail (Watts and Stenner 2012). Q methodology is designed to support the identification and disaggregation of "currently predominant social viewpoints and knowledge structures relative to a chosen subject matter" among a relatively small group of respondents (Watts and Stenner 2012: 42). Thus, the methodology "supports an understanding of the detailed composition of positions, making it suitable ... to understand the perspectives of conservation professionals" (Sandbrook et al. 2013b: 235).

Q methodology begins by requiring respondents to arrange statements drawn from a literature review of the subject onto a grid, as shown in Fig. 6.1. These statements are known as the 'Q set'. We used the same 34 statement Q set developed and used in the ICCB study to enable comparison

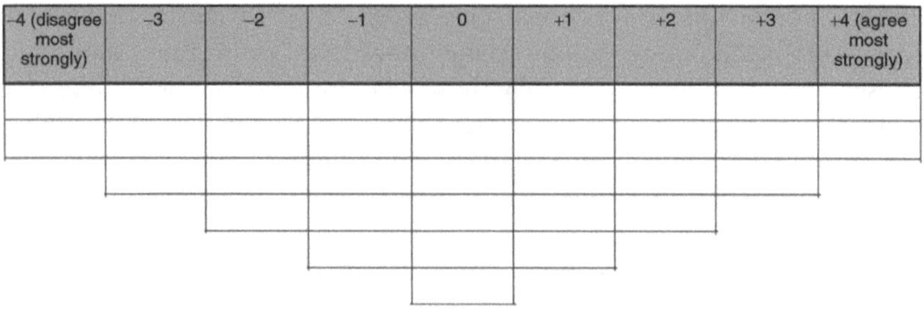

−4 (disagree most strongly)	−3	−2	−1	0	+1	+2	+3	+4 (agree most strongly)

Fig. 6.1 Design of the Q methodology grid used

between the Cambridge and the ICCB data (Table 6.1). This Q set was originally developed by Sandbrook et al. (2013b) using a literature review and the authors' extensive interactions with conservation practitioners and organisations. It was designed to include statements across a continuum of perspectives on market-based approaches to conservation, and it was piloted with two respondents to ensure a balanced sample of statements across a range of perspectives (Sandbrook et al. 2013b).

The Q set of statements included statements about "ethics, pragmatism, ideology and local impacts" as well as a range of perspectives on market-based conservation (Sandbrook et al. 2013b: 11). To be appropriate for a Q study the statements in the Q set needed to be familiar to all respondents and to cover the range of views present within the respondent community.[5] Based on the authors' knowledge of the Cambridge conservation community we felt that the statements developed for the ICCB study met this criterion.

Participants

This Q survey was conducted with employees of conservation organisations located in and around the city of Cambridge. Participants included both those who worked for large international conservation organisations and smaller and for more locally focussed conservation organisations. All the organisations from which staff members were interviewed were involved to some extent in market-based conservation activities. Although the majority of participants were senior employees at the top of their organisational hierarchies, two respondents were mid-career professionals. Eleven were male and six were female. All but two participants were citizens of Organisation for Economic Co-operation and Development (OECD) countries.

Table 6.1 Factor comparison between the Cambridge and ICCB studies

Statement	*Cambridge Factor One*		*ICCB Factor One*		*Cambridge Factor Two*		*ICCB Factor Two*		*Cambridge Factor Three*	
	Z-Score	*Ranking*	*Z-Score*	*Ranking*	*Z-Score*	*Ranking*	*Z-Score*	*Ranking*	*Z-Score*	*Ranking*
1 Markets provide a new source of funding	0.855	2	1.512	3	1.139	2	0.686	−1	0.428	1
2 Markets provide a large source of funding	0.689	1	0.889	2	−0.831	−2	−0.45	−1	−0.391	−1
3 Markets provide a sustainable source of funding	0.684	1	0.785	1	−0.891	−3	−0.387	−1	−0.848	−2
4 Sufficient funds without turning to markets	−1.434	−3	−1.055	−3	−0.791	−2	−0.25	−1	0.305	1
5 Markets most efficient for allocating resources	−0.483	−1	−0.148	0	−0.91	−3	−1.214	−2	0.239	1
6 Markets preferable as conditional on performance	−0.125	0	0.138	0	−0.577	−1	−0.448	−1	−1.116	−3
7 Markets most effective when directly linked to conservation	1.081	3	1.541	4	1.527	3	1.304	3	−0.289	−1
8 Conservation should embrace market, not fight against it	0.885	2	0.508	1	−0.295	0	−1.017	−2	−0.189	−1
9 Biodiversity loss primarily driven by market capitalism	0.823	2	0.083	0	1.59	3	1.003	3	−0.441	−1
10 Biodiversity that cannot survive in the marketplace is not worth conserving	−2.242	−4	−2.011	−4	−2.058	−4	−2.157	−4	−2.174	−4
11 Markets too unpredictable for conservation purposes	−0.208	−1	−0.8	−2	0.719	2	0.758	2	0.318	1
12 Pragmatism not good reason to risk markets	−0.447	−1	−0.714	−1	−0.699	−1	0.828	2	−0.667	−2

(*continued*)

Table 6.1 (continued)

Statement	*Cambridge Factor One*		*ICCB Factor One*		*Cambridge Factor Two*		*ICCB Factor Two*		*Cambridge Factor Three*	
	Z-Score	*Ranking*	*Z-Score*	*Ranking*	*Z-Score*	*Ranking*	*Z-Score*	*Ranking*	*Z-Score*	*Ranking*
13 Decision makers understand monetary values	0.931	3	0.878	1	−0.496	−1	−1	−2	1.202	3
14 Opponents of markets are not living in the real world	0.808	1	−0.998	−2	−0.319	0	−1.518	−3	−0.037	0
15 Private sector partnership undermines conservation	−1.859	−4	−1.518	−3	−0.865	−2	0.492	1	−1.622	−3
16 No difference between commodity and ES markets	−1.201	−3	−0.97	−2	−1.202	−3	−0.775	−2	−1.355	−3
17 Market restructuring cannot deliver conservation	−1.2	−3	−1.159	−3	0.468	1	0.708	2	−0.152	0
18 Nothing really new about market-based approach	−0.314	−1	−0.583	−1	−0.752	−2	−0.058	0	0.449	2
19 By engaging in markets, actors find beneficial outcomes	0.796	1	1.039	2	0.219	1	0.232	0	0.811	2
20 Conservation market expansion has nothing to do with neoliberalism	−0.191	0	−0.176	0	−0.75	−1	−0.701	−1	−0.189	0
21 Conservation organisations should promote the economic valuation of nature	0.601	1	1.084	2	−0.667	−1	−1.425	−3	0	0
22 Conservation organisations should not support commodification of nature	−1.036	−2	−0.964	−2	0.158	0	1.482	4	−0.173	0
23 Markets cannot handle the unpredictable properties of ecosystems	0.107	0	−0.428	−1	1.156	2	1.4	3	1.268	3

24 Need more evidence on the impacts of market conservation	0.369	0	0.775	1	1.692	4	0.556	1	−0.515	−2
25 Putting a price on nature does not detract from other values	1.856	4	1.265	3	−0.345	0	−1.519	−4	0.231	0
26 Conservation choices ethical and political, not solely economic	1.45	4	1.554	4	2.144	4	1.874	4	1.103	2
27 Artificial substitutes may be more competitive than nature	0.516	0	0.155	0	1.218	3	0.225	0	1.898	4
28 Markets have negative impacts where limited exposure of them	−0.107	0	0.158	1	0.465	1	0.43	1	1.173	2
29 Market conservation increases inequality in local communities	−0.517	−1	0.104	0	0.753	2	0.426	0	0.297	1
30 Market transactions are voluntary, so cannot be exploitation	−0.926	−2	−1.532	−4	−1.425	−4	−1.287	−3	−1.783	−4
31 Markets provide livelihood opportunities for the poor	0.878	2	0.908	2	0.394	1	0.2	0	1.326	3
32 Markets deny poor people access to natural resources	−0.857	−2	−0.709	−1	−0.13	0	0.553	1	−0.906	−2
33 Market conservation creates local conservation incentives	1.014	3	1.143	3	0.619	1	0.134	0	2.022	4
34 Private sector partners constrain criticism of markets	−1.198	−2	−0.755	1	−0.26	0	0.916	2	−0.223	−1

Data Collection: The Interview

Data were collected from semistructured, face-to-face interviews with participants in quiet locations away from other people. At the start of the interviews, participants were promised anonymity and asked to share their personal views, as opposed to the views of their organisation. After briefly explaining the project and the Q methodology, respondents were asked to complete the Q survey, in which they sorted the set of 34 printed statements onto a standard distribution grid (see Fig. 6.1). Participants sorted statements over a relative range, from "strongly disagree" (−4) on one side of the grid to "strongly agree" (+4) on the other. This distribution grid required respondents to rank statements relative to other statements, indicating which statements they believed were the most important. Statements were shuffled so that they were presented in a different order to each respondent.

As the sorting took place, respondents were encouraged to explain the rationale for their choices of placement for each statement. In cases where respondents had questions about a statement, a limited explanation of its meaning was given in a way that sought not to introduce bias. After the survey was finished, respondents were asked to explain their reasoning behind statements they chose for the two extremes and for the statements in the middle of the distribution grid. This qualitative data helped in understanding the meaning and significance of participants' choices beyond each participant's sort. We took notes during the interviews as respondents explained their decisions for ranking certain statements, including verbatim quotes for the qualitative component of the data collection. These statements and quotes were then used to interpret and corroborate the results.

Even though participants were encouraged to follow the normal distribution of the grid, eight participants did not arrange their grid like this. Whereas having participants respond within the normal distribution is a practical way of encouraging them to prioritise statements relative to others, it is not essential for use of the method, as even unequally distributed statements reveal their relative level of agreement about each statement in relationship to the others (Brown 1980; Watts and Stenner 2012).

Q Data Analysis

The results of the Q survey, called Q sorts, were input into the PQMethod Software, which is specifically designed for Q methodological analysis. The software analysis requires three statistical steps: (1) correlation, (2) factor analysis and (3) standard factor score computation (Watts and Stenner 2012). Correlation measures the degree of agreement between any two Q sorts and denotes their similarity in a correlation matrix. Factor analysis then searches for patterns of association between the measured variables in the matrix and reduces them into a small number of highly correlated viewpoints, called 'factors' (Watts and Stenner 2012). These factors then undergo 'rotation', which refers to the software plotting identified factors on a three-dimensional correlation matrix and rotating (e.g., moving or adjusting) the matrix to identify and eliminate sorts that are significantly associated onto more than one factor, therefore distinctly defining no factor. This analysis results in each factor identifying a subset of respondents within a Q study who rank-ordered the statements in a similar way, displaying similar perspectives about the provided statements.

For the Cambridge study, we used the software to rotate two, three and four factors and looked at the results. Ultimately, we decided to use and interpret the three-factor result, using two common decision-making criteria (Watts and Stenner 2012). First, we applied the Kaiser-Guttman criterion, which states that the Eigenvalue of a factor should be greater or equal to 1.00 (Watts and Stenner 2012). Second, we accepted Factors One, Two, and Three because they were the only factors that had two or more Q sorts that loaded significantly on each, a common criterion for factor selection (Watts and Stenner 2012).

During the rotation process, the PQMethod 'flags' specific Q sorts that are representative of particular factors. It then generates a 'typical' Q sort, which represents an ideal-type composite version of all Q sorts flagged for that factor. Data interpretation is then done by examining these ideal-type Q sorts alongside the qualitative data. Although not all Q studies include qualitative data, doing so allows for richer interpretation of the results and confirmation that the factor interpretation fits the views expressed by respondents during their interviews.

Narrative descriptions are then written to explain the perspectives defined by each factor. These descriptions, supported by direct quotations from respondents, are presented as results. It should be noted that the interpretation of Q results is a somewhat subjective process (Eden et al. 2005), and comparing factors between Q studies is still experimental, though some published studies have done so (e.g., Tuler and Webler 2009). We chose to do a direct, side-by-side comparison of the idealised factors of both the ICCB and Cambridge datasets to identify points of similarity and difference.

Results

Points of Consensus Among All Factors from Both Studies

All respondents from both the ICCB and Cambridge studies held a limited set of core beliefs, as evidenced by their similar ranking of five statements in the ideal-type sorts, suggesting a common recognition of some limitations of markets as conservation tools in both theory and practise (see Table 6.1). They collectively believed that biodiversity is worth conserving for its multiple values (statement 10), and that conservation choices are ethical and political, not solely economic (statement 26). Respondents from both studies recognised a fundamental difference between markets for traditional commodities and markets for ecosystem services (statement 16). They also recognised the potential for adverse consequences in the use of markets in conservation, including the possibility for exploitation (statement 30).

There was one statement that respondents from both studies did not express a strong opinion on, with relatively neutral responses. This statement was: "the expansion of market-based conservation has nothing to do with neoliberalism" (statement 20). Most Cambridge respondents (16 out of 17) asked (during their Q sort interview) for a definition of neoliberalism when they encountered this statement. Sandbrook et al. also noted in their ICCB study that there was a lack of familiarity among their respondents with this term (Sandbrook et al. 2013b).[6] When asked, interviewers from both studies provided participants with a general definition of neoliberalism that sought not to introduce a bias; however, given respondents' relative unfamiliarity with the term, it is not surprising that this statement did not attract strong views.

Indeed, it is interesting to note that the lack of recognition of this term, which is fairly dominant in the critical literature on markets and conservation, does lend weight to the argument that mainstream conservation practitioners and researchers are not engaging with these critiques (Sandbrook et al. 2013a). This raises the question of how to engage practising conservationists with the latest critical social science literature on conservation, and how social scientists can better communicate and disseminate their research to conservation professionals.

Factor One: Stability Across Both Studies

There was a high level of similarity between Factor One in each study. Respondents associated with this factor expressed support and enthusiasm for the use of market-based interventions in conservation. In comparing Factor One between the two studies, 13 statements were ranked exactly the same on the distributional grid, 17 statements were ranked within one integer of difference, and only four statements (out of a total of 34) were ranked by more than one integer of difference (see Table 6.1). Seven of the 12 respondents from the ICCB study were associated with Factor One, including two senior employees of large international conservation organisations, one government advisor and four academics, of whom two were conservation scientists and two were economists. Likewise, 7 of the 17 respondents from the Cambridge study were associated with Factor One, including four senior employees from large international conservation organisations, a senior employee from a mid-sized international conservation organisation, a senior employee of a local conservation organisation, and a senior employee of a national conservation organisation.

The collective group of respondents that were associated with Factor One were the least sceptical of any group about potential negative effects of markets in conservation. In terms of impacts on local people, respondents associated with Factor One saw little downside to engaging with market-based conservation. They disagreed that market-based conservation denies poor people access to natural resources (statement 32) and with the notion that market-based conservation increases inequality (statement 29). Instead, they saw potential for market-based conservation to create local incentives to support conservation (statement 33) and provide livelihood opportunities for the poor (statement 31). In terms of conservation outcomes, respondents associated with Factor One were indifferent to the idea that "there is a risk that in a market, artificial substitutes may become more competitive than nature at providing services" (statement 27). They were somewhat indifferent to the statement: "Markets have no way of dealing with unpredictable properties of ecosystems, and this makes them dangerous for conservation" (statement 23) and strongly disagreed with the idea that "conservation partnerships with the private sector are undermining conservation outcomes" (statement 15). This group saw little downside in turning to markets.

Unlike respondents to other factors, respondents associated with Factor One disagreed with the notion that markets are too unpredictable to be used for conservation (statement 11). More strongly than respondents associated with other factors, these participants believed that "putting a price on nature does not detract from all the other reasons to value it" (statement 25).

Respondents associated with Factor One in both studies saw the use of markets in conservation as a realistic and necessary tool. They distinctively believed that markets provide both a large (statement 2) and sustainable (statement 3) source of funding for conservation, and clearly believed that sufficient funding to reverse biodiversity loss could not be raised through any method other than markets (statement 4). Respondents associated with this factor held a strong belief that markets can be restructured sufficiently to deliver conservation outcomes (statement 17) and saw markets as potentially helpful in delivering conservation outcomes. They were also the only group to agree with the statements: "Conservationists should embrace market-based capitalism, not fight against it" (statement 8) and "Conservation organisations should promote the economic valuation of nature" (statement 21).

Respondent 22[7] (from the Cambridge study) saw the framing of biodiversity in economic terms as beneficial, explaining that "the conservation community has struggled to express the value of ecosystems. And what we're seeing is that the economic camp is helping lead us into another way of valuing nature". Respondent 15 said: "[I]f we turn our back on monetising nature, we are missing a huge opportunity to embed conservation into our society". Likewise, Respondent 28 said:

[W]e used to be combative and confrontational, presenting to the rest of the world capitalism as the cause of the decline in biodiversity. Now we are moving into a much more mature frame of mind that says collaboration. Let's try to solve these problems together. Let's take what money, wealth, and capitalism can do at face value and help it do the right thing to make the world a better place.

Only four statements from the comparative Factor One sorts were ranked with more than one integer of difference. The Cambridge Factor One respondents agreed more firmly with the statements: "Biodiversity loss is primarily driven by market-based capitalism" (statement 9: Cambridge study, +2; ICCB, 0) and "Decision makers understand monetary values, so conservation should be framed in these terms" (statement 13: Cambridge study, +3; ICCB, +1). In the case of statement 30, "market-based conservation transactions are voluntary, so there is no possibility for exploitation", the ICCB study participants disagreed more strongly (Cambridge study, −2; ICCB, −4).

The largest amount of disparity between Factor One respondents from the two studies was their ranking of statement 14: “Those who oppose market-based conservation are not living in the real world” (Cambridge study, +1; ICCB, −2). This statement was met with verbal scepticism by many of the respondents, some of whom questioned the validity of such a statement in a research setting, despite being reminded that Q methodological research is based on understanding subjective opinions. We felt the statement was perhaps too ambiguous for participants to interpret—many said “it depends what you mean by the real world”; this may have been the reason for such a disparity in ranking of this statement between the Factor One-associated respondents from the two studies.

Various Sceptical Perspectives Within the Remaining Factors

The level of stability between the Factor One idealised sorts was not seen when comparing the remaining factors of the ICCB and Cambridge studies. Even though the remaining factors were distinguished from Factor One on several key points, each portrayed differently nuanced variations of caution and scepticism toward the use of markets in conservation.

Collectively, respondents associated with the remaining factors were not swayed by the rationale that markets provide large (statement 2) or sustainable (statement 3) sources of funding for conservation. Instead, they rejected this funding rationale. Respondent 26 (Cambridge Factor Two) said:

> I don’t think that markets are providing a large source of funding for the right kinds of conservation, ... and the jury is still out on whether markets provide a sustainable source of funding. Sufficient funding hasn’t happened so far [for conservation], so why should it now?

Respondent 14 corroborated this point saying, “no evidence will support that markets provide a sustainable source of funding for conservation. We are miles away from that”.

Those associated with the remaining factors differed from Factor One in that they gave a positive ranking for the statement: “Markets are too unpredictable to be used for conservation purposes” (statement 11). These respondents felt that because markets have no way of dealing with the unpredictable properties of ecosystems, it makes them dangerous for conservation (statement 23), while respondents associated with Factor One from both studies were rather indifferent to this concern. Respondent 23 (Cambridge Factor Three) explained:

> We tend to place values on ecosystems and species based on what is predictable. But we know that a lot of ecosystem processes can be unpredictable. Markets cannot handle things that are unpredictable. We cannot totally depend on markets, because we need mechanisms for dealing with surprises.

Beyond these points of collective disagreement with Factor One, the remaining positions continued to be unique in character across both studies. Each expressed a different, nuanced perspective of caution and scepticism toward the use of markets in conservation. Sandbrook et al. defined the ICCB study's Factor Two as having "ideological scepticism of the underlying rationale for market-based conservation" (2013b). Details of Factors Two and Three in the Cambridge study are given in the following sections.

Cambridge Factor Two: Evidence-Oriented Market Sceptics

Respondents associated with the Cambridge study's Factor Two were sceptical of markets in conservation based on a perceived need for more evidence. Six respondents were flagged for the Cambridge Factor Two. This group included two senior employees from a large international conservation organisation, a senior and mid-level respondent from two large, international conservation organisations, one senior respondent from a national conservation organisation, and a senior employee of a small international conservation organisation.

Respondents associated with the Cambridge Factor Two believed strongly and distinctly that "more evidence is needed on the impacts of market-based conservation before we go too far" (statement 24). Respondent 26 explained that "a lot of people are announcing that partnerships with the private sector are the way forward in conservation, but there is not a lot of evidence behind it". Respondents soundly agreed that "biodiversity loss is primarily driven by market-based capitalism" (statement 9) and were sceptical that markets could be restructured sufficiently to deliver conservation outcomes (statement 17) They worried about the unpredictability of markets (statement 11) and were concerned that "there is a risk that in a market, artificial substitutes may become more competitive than nature at providing services" (statement 27).

Like the ICCB Factor Two, those associated with the Cambridge Factor Two disagreed somewhat with the rationale that "decision makers understand monetary values, so conservation should be framed in those terms" (statement 13). Instead, respondents felt that decision makers understand more than just economic arguments, and that conservationists do not necessarily have to cater to economic rationales for their arguments to be recognised. Respondent 16 said: "I don't think it's true that decision makers are only moved by the economic argument. Decision makers are moved by all sorts of reasons. It's a myth that they are only swayed by markets". Respondents did not think efficiency was a good reason to engage with markets in conservation (statement 5). Respondent 16 said: 'I don't think the concept of efficiency applies to biodiversity'.

Cambridge Factor Three: Social Outcome-Focused Realists

The ICCB study yielded two factors, while the Cambridge study yielded three factors. Respondents associated with Cambridge Factor Three saw the potential for the use of markets in conservation to benefit local people but were sceptical about the ability of markets to deliver biodiversity conservation outcomes. Three respondents were associated with this factor, including a senior employee of a small international conservation organisation, a senior employee of a small national conservation organisation, and a mid-level employee of a large international conservation organisation (its senior employees interviewed were flagged to either Factor One or Factor Two). When we ran a two-factor solution for the Cambridge study and compared it to the final three-factor solution, one of these respondents was flagged for Factor One in the two-factor solution, whereas the other two were flagged for Factor Two in a two-factor solution. This suggests that Factor Three is made up of hybrid viewpoints from both Factors One and Two from the Cambridge study. Indeed, Factor Three reveals a pragmatism for using markets that was reflected in both the ICCB and Cambridge studies' Factor One, together with a scepticism toward the ability of markets to deliver certain outcomes, which was reflected in the Cambridge Factor Two.

Respondents associated with Factor Three saw utility in using market-based conservation to create incentives for local people (statement 33) and believed that "market-based conservation provides livelihood opportunities for the rural poor" (statement 31). Although respondents associated with this factor saw the potential of markets to deliver positive social outcomes, they recognised that social outcomes are context-specific and dependent on how market-based conservation initiatives are implemented. For example, unlike Factor One respondents, these respondents saw the possibility of markets having negative social impacts in places with limited experience with the market economy (statement 28). They also believed that even though market-based conservation is voluntary, this does not negate the possibility for it to be exploitive (statement 30). Respondent 19 said that "markets could go both ways. [They] could increase or decrease inequality"; respondent 23 agreed:

> [F]or a market to operate successfully, all actors need access to all information. Unfortunately, many people in local communities have limited information. If this is the case, if you introduce a market system, the local people end up losing out. But if the distribution of revenue is well done, the issue of inequality should not arise.

Even though those associated with this factor saw the potential for positive social impact from market-based conservation, respondents associated with it were less optimistic about the ability of markets to deliver biodiversity conservation outcomes. Like Factor Two, respondents linked to Factor Three strongly agreed that "there is a risk that in a market, artificial substitutes may become more competitive than nature at providing services" (statement 27). Respondent 27 said: "[S]uppose something comes along, product B, that's made very cheaply. Then the [ecosystem service] market would slump and we'd lose all traction. Markets have unpredictable slumps that are scary". Respondent 19 said: "I think markets are very risky, as we've experienced with the economic downturn. You should never rely on markets to solve the biodiversity problem".

Respondents associated with this factor seemed to be looking beyond markets for other means to conserve biodiversity. This was the only group that strongly disagreed with the idea that "market-based conservation is preferable to other forms because it is conditional on performance" (statement 6). This factor was also the only more sceptical factor with respondents who disagreed with the idea that "markets are most effective for conservation when they are directly linked to the delivery of conservation outcomes" (statement 7). These responses suggest that Factor Three participants do not believe that market-based conservation is the most preferable or effective means of doing conservation, either despite or because this form of conservation is sometimes linked to the delivery of conservation outcomes. Their disagreement with statements 6 and 7 implies that this group is sceptical of the ability of markets to deliver on conservation outcomes, and believe that other methods of conservation may be more effective and better at ensuring biodiversity outcomes.

Interestingly, the wariness of those associated with Factor Three about the ability of markets to deliver conservation outcomes does not come from a feeling that there needs to be more evidence, as the Factor Two perspective suggested. On the contrary, Factor Three was the only factor across both studies where respondents disagreed with "we need more evidence on the impacts of market-based conservation before we go too far" (statement 24). Respondent 19 said, "we could keep going for more evidence. But in the end, we just need to do something".

Factor Three respondents did believe, along with Factor One, that "decision makers understand monetary values, so conservation should be framed in those terms" (statement 13). This statement postulates how decision makers behave, and although respondents agreed with this description, they did not agree with this statement on normative terms or approve of its reality. Respondent 23 explained that "politicians would prefer that conservation is conveyed in messages that are easier to understand. When you convert ecosystem services into [economic] values, this tends to convey a stronger message than if you just asked them to put money into something that they don't understand". Likewise, respondent 19 said: "I agree quite strongly that decision makers understand monetary values. I don't think they know what biodiversity actually means". However, signaling their disapproval with this reality, Factor Three is the only factor in which respondents convey disagreement with the idea that conservationists should embrace market-based capitalism (statement 8).

Discussion

This comparative study found a pro-market perspective that was consistent across both study groups, and three fragmented and more critical perspectives that differed across the study groups. Although some positive perspectives in each study group might have been expected given the rise of MBIs across the conservation community, the consistency of the statements that loaded onto the pro-markets Factor One across both studies is remarkable given that Q methodology is a sophisticated tool capable of uncovering highly detailed and nuanced subjective value positions (Watts and Stenner 2012). This result lends support to the suggestion that, although sampled in very different geographies, the individuals aligned with Factor One in both studies were in fact members of a single transnational epistemic community, with a shared way of thinking about the role of markets in conservation.

In contrast to the consistency of Factor One, the other factors identified were distinct in character both within and across the two studies: Factor Two from the ICCB study was different in character than the Cambridge Factor Two, and the third factor that emerged from the Cambridge study offered an additional critical perspective toward the use of markets in conservation. This tension of perspectives over the use of MBIs in conservation existed not only between members of different conservation organisations, but between staff members within the same organisation. This occurred even within organisations that have advocated and adopted market-oriented conservation activities. The lack of consistency in these factors across the two study groups suggests that the individuals aligned with these factors belong to a more fragmented discursive community, in contrast to the result for Factor One in both studies.

Why might we find a shared, supportive perspective on the role of markets in conservation among conservationists sampled at different times, on opposite sides of the globe? One possibility is that those holding these views have been exposed to a consistent and influential set of global communications that have promoted this view, such as the 2010 'The Economics of Ecosystems and Biodiversity' study (TEEB 2010), or the Convention on Biological Diversity's 'Strategic Plan for Biodiversity 2011–2020' (CBD 2010). To our knowledge, no equivalent critical communication on the role of markets in conservation have been as influential, or similarly promoted, among conservationists.

A second possibility is that some of the individuals we sampled actively participate in transnational networks and events within which such ideas are created and circulated.[8] Participants in such networks might be expected to be more senior within their organisations because it is usually senior staff who attend the conferences and conventions central to the operations of such networks (Campbell 2010; MacDonald 2010a; Büscher 2014; Campbell et al. 2014). To investigate this possibility within our study's data, we examined the relationship between each respondent's relative seniority within each's conservation organisation, the relative size of the organisation, and the individual's association with each factor. One pattern that emerged was that senior-level respondents from large, internationally focussed conservation organisations were associated with the pro-market Factor One (Table 6.2). Likewise, respondents who loaded

Table 6.2 Respondents by seniority, organisation size and factor

Study	*Number*	*Size of organisation*	*Respondent's rank in organisation*	*Respondent's factor flag*
ICCB	1	University	Mid-level	1
ICCB	2	Large/Int'l	Senior	1
ICCB	3	Large/Int'l	Mid-level	2
ICCB	4	University	Mid-level	1
ICCB	5	Large/Int'l	Mid-level	2
ICCB	6	Large/Int'l	Mid-level	2
ICCB	7	Large/Int'l	Senior	1
ICCB	8	n/a	n/a	1
ICCB	9	Large/Int'l	Mid-level	2
ICCB	10	Small/Local	Mid-level	2
ICCB	11	University	Mid-level	1
ICCB	12	University	Senior	1
Cambridge	13	Large/Int'l	Senior	1
Cambridge	14	Large/Int'l	Senior	2
Cambridge	15	Large/Int'l	Senior	1
Cambridge	16	Large/Int'l	Senior	2
Cambridge	17	Large/Int'l	Senior	1
Cambridge	18	Large/Int'l	Senior	2
Cambridge	19	Large/Int'l	Mid-level	3
Cambridge	20	Large/Int'l	Mid-level	2
Cambridge	21	Large/Int'l	Senior	1
Cambridge	22	Medium/Int'l	Senior	1
Cambridge	23	Small/Int'l	Senior	3
Cambridge	24	Large/Int'l	Senior	n/a—did not flag
Cambridge	25	Small/Local	Senior	1
Cambridge	26	Small/Int'l	Senior	2
Cambridge	27	Small/Local	Senior	3
Cambridge	28	Medium/National	Senior	1
Cambridge	29	Medium/National	Senior	2

onto the more sceptical factors in both studies tended to hold mid-level positions or work for smaller, national or more locally focussed conservation organisations. This pattern was apparent even among employees of the same conservation organisations within our sample.

Given the scope of this study, the authors acknowledge that we are unable to say whether this retrospective observation on seniority exists outside of our sample. In particular, the Q methodology study design does not allow inferential conclusions to be drawn, regarding for instance the relationship between association with factors and the demographic characteristics of the sample. These initial empirical data, however, do provide some support for Holmes's proposed epistemic community (2011)—a 'transnational policy elite'—promoting 'fast' neoliberal conservation policy (after Peck and Theodore 2010b). The expression of a nearly identical perspective about market-based approaches in conservation across individuals in senior positions within two distinct conservation networks suggests that there is a consistent and apparently durable pro-market perspective that exists and persists among elite conservation professionals from diverse geographies. This might be sharpened into 'alignment' (Wenger 2000; Peck and Theodore 2010a) at various points of overlap within the conservation community's transnational policy networks, and at events (e.g., international conferences) where such transnational elites come together (MacDonald 2010a).

In contrast to Factor One, the more sceptical perspectives found in Factor Two of the ICCB study, and the distinct Factors Two and Three from the Cambridge study, were nonuniform and lacked the alignment seen in the pro-market Factor One. This points to the fragmented nature of dissent within the conservation community among those who are more sceptical toward the dominant thinking about neoliberal conservation. These more sceptical factors, primarily voiced in our small and non-representative sample by mid-level conservation professionals and those in smaller, nationally or regionally focussed conservation organisations, provide some empirical substance to the suggested fractured nature of less dominant ideas, which lack the alignment of an actively promoted perspective.

Further, in examining the perspectives represented by these less dominant factors, it should be noted that each sceptical factor was sceptical primarily on the grounds of inconclusive evidence (Cambridge Factor Two), on-the-ground pragmatism (Cambridge Factor Three) or cautious pragmatism merged with a scepticism about whether markets are too unpredictable and problematic to be applied to conservation (ICCB study, Factor Two). In other words, despite being warier of market-based approaches, all the more sceptical factors represented viewpoints of support for the deployment of market-based policy under certain conditions.

Thus, even within the sceptical factors across both studies, we found no factor representing a critique that comes close to the perspective of critical social scientists studying the use of MBIs, despite the Cambridge respondents being part of a geographically co-located conservation network that includes such critical scholars. Indeed, the lack of familiarity with the term 'neoliberalism' among all participants corroborates this point. It would be very interesting to conduct similar research among the epistemic community of critical social science scholars writing about neoliberal conservation to establish whether a globally shared *critical* perspective on the role of markets in conservation exists among them, along the same lines as the consistent Factor One we identified among conservationists.[9]

The absence of strongly critical viewpoints among the conservationists in our study groups could be the outcome of what Büscher et al. (2012) have called the 'disciplining of dissent' within neoliberal conservation ideology. This could happen in two ways. First, despite the study participants being asked to express their personal viewpoints, they may have been reluctant to express critical views that are counter to the pro-market perspective that has so much momentum. Second, people with critical views may be absent from mainstream conservation research and practise organisations because they are not hired in the first place.

Concluding Remarks

This comparative study identified a strikingly consistent pro-market perspective surrounding the use of MBIs in conservation among two groups of conservation professionals, in two different contexts where conservation professionals convene. The similarity of this pro-market perspective corroborated the observation that the adoption of a particular neoliberal, market-led approach to conservation exists and persists within the broader biodiversity conservation community. The evidence also supports a growing body of research that suggests that pro-market and neoliberal approaches to conservation are embedded within the thought processes of some decision makers and staff of conservation organisations. This finding also lends some support to the proposed existence of a transnational conservation elite (Holmes 2011) that supports these developments. Second, this research corroborates earlier findings that pro-market perspectives in conservation are not altogether uncontested. Our comparative results show that the consistency of pro-market perspectives was matched by fragmented sceptical perspectives among other conservation professionals. The additional critical perspectives that emerged in the Cambridge study validates the ICCB study's findings that there is some tension in the way that conservationists think about markets (Sandbrook et al. 2013b), as well as a plurality of perspectives in the ways that conservationists value nature (Sandbrook et al. 2011).

The Q methodology offers the ability to quantitatively study subjectivity, allowing detailed comparison of perspectives. Although taking a comparative approach to Q studies is relatively new, this approach allows for further exploration into the perspectives of conservation actors and professionals across diverse conservation organisations and networks. This study, and future studies in this area, could be supplemented by other Q study designs, like the one followed by Rastogi et al. (2013), that use online surveys to measure the popular approval of viewpoints identified by Q. Further research could continue to explore the indications we have reported here regarding the influence of seniority and the extent to which leaders within large, internationally focussed conservation organisations circulate pro-market perspectives.

Notes

1. Cambridge Conservation Initiative: Transforming the landscape of biodiversity conservation. http://www.conservation.cam.ac.uk/. Accessed 12 February 2014.
2. Cambridge Conservation Forum. http://www.cambridgeconservationforum.org.uk/. Accessed 12 February 2014.
3. Cambridge Conservation Initiative: Transforming the landscape of biodiversity conservation. http://www.conservation.cam.ac.uk/. Accessed 12 February 2014.
4. University of Cambridge Conservation Research Institute. http://research-institute.conservation.cam.ac.uk/. Accessed 12 February 2014.
5. Even though we thought that all the statements would be familiar, given Sandbrook et al.'s 2013 testing of the Q sort in the original study, we actually found this not to be the case. One statement, which focused on the term 'neoliberalism' was unfamiliar to the Cambridge study participants. Indeed, this was a discovery of an ill-designed study question that ended up soliciting a response that was illuminating (see our discussion in the Results section of the chapter).
6. Although the term 'neoliberal' is not commonly invoked in biodiversity conservation conferences or email lists, promoters of market-based conservation "effectively associate the central elements of neoliberal conservation without using a single related word" (Igoe and Brockington 2007: 435). Büscher and Dressler (2007) call this a 'discursive blur'.
7. Note that all respondents from each study were given a unique number, as shown in Table 6.2, so as to not confuse respondents across both studies. Respondents 1–12 are from the ICCB study, while respondents 13–29 are from the Cambridge study.

8. Such network events include the IUCN's World Conservation Congress and the Society for Conservation Biology meetings (see MacDonald 2010a).
9. Such research could potentially be conducted at critical social science events such as "Grabbing Green" or the "STEPS: Resource Politics" conferences.

References

Arsel, M., and B. Büscher. 2012. Nature™ Inc.: Changes and Continuities in Neoliberal Conservation and Market-Based Environmental Policy. *Development and Change* 43: 53–78.

Balmford, A., and T. Whitten. 2003. Who Should Pay for Tropical Conservation, and How Could the Costs Be Met? *Oryx* 37 (2): 238–250.

Brannstrom, C. 2011. A Q-Method Analysis of Environmental Governance Discourses in Brazil's Northeastern Soy Frontier. *The Professional Geographer* 63: 531–549.

Brockington, D. 2002. *Fortress Conservation: The Preservation of the Mkomazi Game Reserve, Tanzania*. Oxford: International African Institute.

Brockington, D., and R. Duffy. 2011. *Capitalism and Conservation*. Chichester: Wiley Blackwell.

Brown, S. 1980. *Political Subjectivity: Applications of Q Methodology in Political Science*. New Haven: Yale University Press.

Büscher, B. 2014. Collaborative Event Ethnography: Between Structural Power and Empirical Nuance? *Global Environmental Politics* 14 (3): 132–138.

Büscher, B., and W. Dressler. 2007. Linking Neoprotectionism and Environmental Governance: On the Rapidly Increasing Tensions Between Actors in the Environment-Development Nexus. *Conservation and Society* 5 (4): 586–611.

Büscher, B., S. Sullivan, K. Neves, J. Igoe, and D. Brockington. 2012. Towards a Synthesized Critique of Neoliberal Biodiversity Conservation. *Capitalism Nature Socialism* 23 (2): 4–30.

Campbell, L.M. 2010. Collaborative Event Ethnography: Conservation and Development Trade-Offs at the Fourth World Conservation Congress. *Conservation and Society* 8 (4): 245–249.

Campbell, L.M., C. Corson, N.J. Gray, K.I. MacDonald, and J.P. Brosius. 2014. Studying Global Environmental Meetings to Understand Global Environmental Governance: Collaborative Event Ethnography at the Tenth Conference of the Parties to the Convention on Biological Diversity. *Global Environmental Politics* 14 (3): 1–20.

Castree, N. 2008. Neoliberalising Nature: The Logics of Deregulation and Reregulation. *Environment and Planning A* 40: 131–152.

Convention on Biological Diversity (CBD). 2010. Strategic Plan for Biodiversity 2011–2020. https://www.cbd.int/sp/. Accessed Feb 2013.

Daily, G. 1997. *Nature's Services: Societal Dependence on Natural Ecosystems.* Washington, DC: Island Press.

Eckert, P. 2006. Communities of Practice. In *Encyclopedia of Language and Linguistics*, ed. K. Brown, vol. 1, 2nd ed., 683–685. London: Elsevier B.L.

Eden, S., A. Donaldson, and G. Walker. 2005. Structuring Subjectivities? Using Q Methodology in Human Geography. *Area* 37 (4): 413–422.

European Commission. 2011. Our Life Insurance, Our Natural Capital: An EU Biodiversity Strategy to 2020. http://ec.europa.eu/environment/nature/biodiversity/comm2006/pdf/EP_resolution_april2012.pdf. Accessed 14 Feb 2013.

Ferraro, P.J. 2001. Global Habitat Protection: Limitations of Development Interventions and a Role for Conservation Performance Payments. *Conservation Biology* 15 (4): 990–1000.

Haas, P.M. 1992. Introduction: Epistemic Communities and International Policy Coordination. *International Organization* 46 (01): 1.

Harvey, D. 2003. *The New Imperialism.* Oxford: Oxford University Press.

———. 2006. *Spaces of Global Capitalism: Towards a Theory of Uneven Geographical Development.* London: Verso.

Heynen, N., and P. Robbins. 2005. The Neoliberalization of Nature: Governance, Privatization, Enclosure and Valuation. *Capitalism Nature Socialism* 16 (1): 5–8.

Holmes, G. 2011. Conservation's Friends in High Places: Neoliberalism, Networks, and the Transnational Conservation Elite. *Global Environmental Politics* 11 (4): 1–21.

Igoe, J., and D. Brockington. 2007. Neoliberal Conservation: A Brief Introduction. *Conservation and Society* 5 (4): 432–449.

Landell-Mills, N., and I.T. Porras. 2002. *Silver Bullet or Fools' Gold? A Global Review of Markets for Forest Environmental Services and Their Impact on the Poor*, Instruments for Sustainable Private Sector Foresty Series. London: International Institute for Environment and Development.

Lave, J., and E. Wenger. 1991. *Situated Learning: Legitimate Peripheral Participation.* Cambridge: Cambridge University Press.

López-i-Gelats, F., J.D. Tàbara, and J. Bartolomé. 2009. The Rural in Dispute: Discourses of Rurality in the Pyrenees. *Geoforum* 40 (4): 602–612.

MacDonald, K.I. 2010a. Business, Biodiversity and New 'Fields' of Conservation: The World Conservation Congress and the Renegotiation of Organisational Order. *Conservation and Society* 8 (4): 268.

———. 2010b. The Devil Is in the (Bio)diversity: Private Sector "Engagement" and the Restructuring of Biodiversity Conservation. *Antipode* 42 (3): 513–550.

Mazur, K.E., and S.T. Asah. 2013. Clarifying Standpoints in the Gray Wolf Recovery Conflict: Procuring Management and Policy Forethought. *Biological Conservation* 167: 79–89.

McAfee, K. 2012. The Contradictory Logic of Global Ecosystem Services Markets. *Development and Change* 43 (1): 105–131.

McCarthy, J., and S. Prudham. 2004. Neoliberal Nature and the Nature of Neoliberalism. *Geoforum* 35 (3): 275–283.

Muradian, R., M. Arsel, L. Pellegrini, F. Adaman, B. Aguilar, B. Agarwal, E. Corbera, et al. 2013. Payments for Ecosystem Services and the Fatal Attraction of Win-Win Solutions. *Conservation Letters* 6 (4): 274–279.

Pagiola, S., A. Arcenas, and G. Platais. 2005. Can Payments for Environmental Services Help Reduce Poverty? An Exploration of the Issues and the Evidence to Date from Latin America. *World Development* 33 (2): 237–253.

Pattanayak, S.K., S. Wunder, and P.J. Ferraro. 2010. Show Me the Money: Do Payments Supply Environmental Services in Developing Countries? *Review of Environmental Economics and Policy* 4 (2): 254–274.

Pearce, D.W., and E. Barbier. 2000. *Blueprint for a Sustainable Economy*. London: Earthscan.

Peck, J. 2011. Global Policy Models, Globalizing Poverty Management: International Convergence or Fast-Policy Integration? *Geography Compass* 5 (4): 165–181.

Peck, J., and N. Theodore. 2010a. Mobilizing Policy: Models, Methods, and Mutations. *Geoforum* 41 (2): 169–174.

———. 2010b. Recombinant Workfare, Across the Americas: Transnationalizing "Fast" Social Policy. *Geoforum* 41 (2): 195–208.

Pirard, R. 2012. Market-Based Instruments for Biodiversity and Ecosystem Services: A Lexicon. *Environmental Science & Policy* 19–20: 59–68.

Rastogi, A., G.M. Hickey, R. Badola, and S.A. Hussain. 2013. Diverging Viewpoints on Tiger Conservation: A Q-Method Study and Survey of Conservation Professionals in India. *Biological Conservation* 161: 182–192.

Robbins, P. 2000. The Practical Politics of Knowing: State Environmental Knowledge and Local Political Economy. *Economic Geography* 76 (2): 126–144.

———. 2006. The Politics of Barstool Biology: Environmental Knowledge and Power in Greater Northern Yellowstone. *Geoforum* 37 (2): 185–199.

Sandbrook, C.G., I.R. Scales, B. Vira, and W.M. Adams. 2011. Value Plurality Among Conservation Professionals. *Conservation Biology* 25 (2): 285–294.

Sandbrook, C.G., W.M. Adams, B. Büscher, and B. Vira. 2013a. Social Research and Biodiversity Conservation. *Conservation Biology*, 1–4.

Sandbrook, C.G., J.A. Fisher, and B. Vira. 2013b. What Do Conservationists Think About Markets? *Geoforum* 50: 232–240.

Sandel, M. 2012. *What Money Can't Buy: The Moral Limits of Markets*. London: Penguin Books.

Schurman, R.A., and D.T. Kelso. 2003. *Engineering Trouble: Biotechnology and Its Discontents*. Berkeley: University of California Press.

Stokstad, E. 2016. Conservation Researchers Get a New Roost in Cambridge. *Science* 351 (6269): 114.

Storm, S. 2009. Capitalism and Climate Change: Can the Invisible Hand Adjust the Natural Thermostat? *Development and Change* 40 (6): 1011–1038.

The Economics of Ecosystems & Biodiversity (TEEB). 2010. Mainstreaming the Economics of Nature: A Synthesis of the Approach, Conclusions and Recommendations of TEEB. http://www.teebweb.org/our-publications/teeb-study-reports/synthesis-report/. Accessed 20 Feb 2013.

Tuler, S., and T. Webler. 2009. Stakeholder Perspectives About Marine Oil Spill Response Objectives: A Comparative Q Study of Four Regions. *Journal of Contingencies and Crisis Management* 17 (2): 95–107.

United Nations Environment Programmes (UNEP). 2011. Towards a Green Economy: Pathways to Sustainable Development and Poverty Eradication. http://www.unep.org/greeneconomy. Accessed 13 Jan 2013.

Vira, B. 2002. Trading with the Enemy? Examining North-South Perspectives in the Climate Change Debate. In *Economics, Ethics, and Environmental Policy*, ed. D.W. Bromley and J. Paavola, 164–180. Oxford: Blackwell Publishing.

———. 2015. Taking Natural Limits Seriously: Implications for Development Studies and the Environment. *Development and Change* 46 (4): 762–776.

Watts, S., and P. Stenner. 2012. *Doing Q Methodological Research: Theory, Mind and Interpretation*. London: Sage Publications.

Wenger, E. 2000. Communities of Practice and Social Learning Systems. *Organization Articles* 7 (2): 225–246.

Wunder, S. 2007. The Efficiency of Payments for Environmental Services in Tropical Conservation. *Conservation Biology* 21 (1): 48–58.

CHAPTER 7

Conservation Jujutsu, or How Conservation NGOs Use Market Forces to Save Nature from Markets in Southern Chile

George Holmes

Introduction

In recent years, an expanding literature on neoliberal conservation has examined the changing nature of conservation governance (Igoe and Brockington 2007; Brockington and Duffy 2010; Arsel and Buscher 2012). In particular, it has identified how conservation nongovernmental organisations (NGOs) (1) have increased in number, size and importance; and (2) how market-based mechanisms increasingly are used to save biodiversity. This chapter examines the relationship between markets and NGOs in the context of neoliberal conservation, focussing on how and why conservationists have engaged with markets to save biodiversity. This engagement has taken various forms, from providing advice to businesses to promoting market-based instruments and approaches, such as payments for ecosystem services and ecotourism, to intervening directly in property ownership—for example, by buying land for conservation (Igoe and Brockington 2007; Robinson 2012).

G. Holmes (✉)
School of Earth and the Environment, University of Leeds, Leeds, UK

P.B. Larsen, D. Brockington (eds.), *The Anthropology of Conservation NGOs*, Palgrave Studies in Anthropology of Sustainability, DOI 10.1007/978-3-319-60579-1_7

The literature has identified a range of opinions within conservation NGOs about why they should engage with markets. At one end of the spectrum some conservationists, particularly those associated with the so-called New Conservation (e.g., Peter Kareiva, former chief scientist at the Nature Conservancy), argue that markets and corporations can be a force for good for conservation, and they are enthusiastic about their potential (Lalasz et al. 2011; Kareiva et al. 2014; see also Wynne-Jones 2012). Not only are they seen to bring forth greater funding for conservation, but also to produce new innovations that allow conservation to harness and redirect market forces to save biodiversity.

Another example of such a position would be that of philanthrocapitalist conservationists, who argue that conservation organisations are the most effective when they reason and act like businesses (Holmes 2012). For example, the Wildlife Conservation Network (WCN), based in Silicon Valley, tries to apply the same techniques of venture capitalist investment used to fund technology start-ups to finance conservation organisations, supporting what it calls conservation entrepreneurs. The WCN argues:

> Like their counterparts in business, successful conservation entrepreneurs are creative, action-oriented and dare to try different strategies. Their passion and commitment to wildlife preservation compels them to undertake conservation projects in areas the mainstream considers too risky, too arduous, or too formidable ... [who] maximise the value of every dollar invested in conservation. (quoted in Holmes 2012)

A different, less enthusiastic position is that engaging with markets is a necessary evil. The argument here is that because we live in a market-based society, where markets are used to determine all kinds of processes in our contemporary lives, then conservationists cannot ignore markets but must engage with them because they are the dominant power force (Robinson 2012; see Holmes 2010). A variation of this position, explored within this chapter, is that conservationists should engage with markets to save nature from markets. Much like the martial art of Jujutsu, in which combatants attempt to use the forward momentum and force of their opponents to eventually destabilise and defeat them, this position might argue that conservationists ultimately should use markets in natural resources to save natural resources from markets.[1]

Understanding why conservation NGOs are engaging with markets is important for two broad reasons. First, it can provide key insights into how neoliberal conservation operates. Within the literature on neoliberal conservation, there are two divergent positions on why conservation and capitalism have become more entangled. Those approaching the issue from a political economy position have argued that this entanglement has emerged because capitalist actors see the environment as a new frontier for extracting surplus capital—conservation is just the newest way of making a profit (Castree 2008a; Büscher et al. 2012).

Others approaching it from the perspective of NGO politics—particularly relationships between NGOs, donors and corporations—have argued that the turn has been led by a wish among NGOs to get access to the right political arenas, money and other tools to undertake their work; that is, the use of markets is another, more powerful way of 'doing' conservation (Corson 2010; Holmes 2010). These approaches and ideas are not incompatible or contradictory, but the combination of broad trends in capitalist accumulation and the decisions made in individual NGO boardrooms and donor organisations that indicate the complex and multifaceted nature of neoliberal conservation.

Second, understanding conservation NGOs can provide key insights into which kinds of policies and approaches emerge in the context of neoliberal conservation, particularly how ideologies and perspectives (e.g., the New Conservation) may translate into actual day-to-day practices and their effects on nature and people. Asking how and why conservation NGOs are engaging with markets requires a consideration of the links between the position and the strategy of an NGO, as decided in its boardroom, and the implementation of these positions on the ground in specific conservation projects. This is a necessarily messy process because the simplified ideas of strategy documents are implemented on the dynamic, heterogeneous and unpredictable society living near a protected area (PA) or ecotourism project.

Neoliberalism, Conservation and NGOs

Neoliberal approaches to conservation have greatly increased the role of markets and NGOs in saving biodiversity. A key feature is that the state has 'rolled back' from direct involvement in saving biodiversity. Instead, it has 'rolled out' new mechanisms and structures that have allowed markets to take a key role in saving biodiversity (Peck and Tickell 2002; Castree 2008b). For example, states are rolling out structures for payments for ecosystem services, biodiversity offsetting and other market-based instruments or programmes in which private sector ecotourism or hunting operators are put at the centre of a conservation strategy. Although market-based instruments and strategies long have had a place in conservation, they have never been so varied, with many new forms (e.g., payments for ecosystem services) emerging in recent decades, nor so common, nor have they been accompanied such triumphalist rhetoric (Igoe and Brockington 2007; Brockington and Duffy 2010).

The NGOs, donors, philanthropists and corporations promote a discourse in which (1) capitalism and conservation are seen as complementary, not contradictory; and (2) market-based solutions can be found that benefit conservation and local people and global, national and local economies, so long as they are designed appropriately (Brockington and Duffy 2010). Here, nature often is understood in economic terms as 'natural capital' but tacitly and implicitly framed through the lens of neoclassical economics (Büscher et al. 2012). An interesting reflection of the lack of deeper reflection on the type of economics being used by conservationists is that, even though the term 'neoliberalism' is commonplace in the social science literature on conservation, it is entirely absent from the lexicon of conservation professionals in Blanchard et al.'s study (2016).

Ultimately, such neoclassical framings are not just rhetorical but also affect the kinds of funding and partnerships that emerge from thinking about nature and economics in a certain way; thus, the framing of the relationship between economics and nature has a real-world effect on the kinds of projects and policies that emerge (Fisher and Brown 2014). This is evident especially in the Chilean data we discuss later, where conservationists and policymakers talk about economics and nature generally, but also where they have particularly narrow ways of understanding the economy, and nature's role in it.

A second key feature of neoliberalism is that NGOs have become larger and more influential. As states have rolled back from day-to-day interventions in biodiversity conservation, they have rolled out mechanisms (or left political vacuums) that allow NGOs to replace the state's role in conservation. At the same time, large international NGOs have successfully positioned themselves with policymakers in government and donor agencies as the only actors capable of conserving biodiversity on a transnational scale, given that endangered species do not respect national boundaries.

As Hrabanski et al. (2013) note, a key part of NGOs' influence is their role as policy entrepreneurs, transferring ideas developed in one place and context to another. They argue that NGOs are particularly strong policy entrepreneurs in countries where the state is weak, leaving a vacuum for

NGOs to provide policy (see also Mbembé 2001; Igoe and Fortwangler 2007; Igoe et al. 2010). It also is possible that they could be especially influential in countries where the state is unwilling, not necessarily unable, to engage in biodiversity conservation, principally highly neoliberal conservation where the state has substantially rolled back.

The net result of this is that conservation NGOs have become more powerful and influential at all levels of conservation in recent decades, from global-scale negotiations to local-scale day-to-day politics of implementing conservation projects, and the methodologies and field sites used by those studying NGOs has changed accordingly. This is not to say that their power is never contested or constrained, but that there is a general trajectory in which their influence has grown in recent years.

Bringing these features together, it is clear that international NGOs are increasingly engaging with market mechanisms, and this may be driving their global spread (Brockington and Scholfield 2010). Although some individual NGOs have longer histories of using markets, there has been an expanding of this type of engagement. Conservation conferences have moved from discussing valuing nature as a means of justifying conservation, such as through ecotourism, to discussing paying for nature, such as through marketised offsets (Brosius and Campbell 2011). Increasingly, NGOs are taking over the running of nominally state-governed PAs, financing and governing them using market tools such as ecotourism (Igoe and Brockington 2007; Holmes 2012).

This chapter considers a more extensive engagement between markets and NGOs, where some are purchasing land themselves to establish protected areas. Several NGOs have moved from seeing ecosystem services as a metaphor to a resource to be monetised. Since the late 2000s, conservation–business partnerships have been advocated and presented in leading conservation conferences as an essential and largely uncontested part of conservation (MacDonald 2010). Dissent within conservationists against neoliberalism is not very visible within many key policy conferences (Büscher et al. 2012). This may because it is an ideology embedded in the dominant structures of conservation (e.g., international agreements, international NGOs and other key players), and the techniques and approaches that dominate—ecotourism for one (Büscher et al. 2012).

A complimentary argument is that while leading 'elite' (Holmes 2011) actors in conservation, the key thinkers and decision makers in NGOs, government and international agencies may be in favour of neoliberal approaches. This view may not be shared by the broader constituency of conservationists outside of this narrow but powerful group. Recent empirical analyses of the varied discourses on markets among conservation professionals provide some support for this idea (Sandbrook et al. 2011; Blanchard et al. 2016). Additionally, it may be that the leadership in conservation NGOs may be suppressing dissent. For example, MacDonald's (2010) study of the International Union for the Conservation of Nature (IUCN) World Parks Congress shows how the IUCN presents its relationship with business as a productive, tension-free collaboration, and it stage manages conferences so that dissenting voices are actively excluded.

Whereas, there may be some dissent within the pages of conservation journals (e.g., the contrast between Soule 2013 and Kareiva 2014) and conferences, these are arenas dominated by academics and few key decision makers attend or read their publications. Furthermore, even though some voices in these debates may express concern over the integration of capitalism and conservation, very few use concepts originating in the social science literature on neoliberalism (e.g., see Blanchard et al. 2016).

The cases in the empirical literature demonstrate several rationales for why NGOs have engaged with markets, corresponding to various points on the scale between being enthusiastic about the benefits of using markets, to engaging with them as a necessary evil to the Jujutsu approach. First, engaging with markets has had the effect of broadening the constituency for conservation. Discourse analysis of conservation professionals has shown that some view the use of market tools and market metaphors as a useful way of accessing key decision makers; of speaking the language of donors and politicians; and, ultimately, of making them consider conservation as a useful and worthwhile activity (Sandbrook et al. 2011; Fisher and Brown 2014; Blanchard et al. 2016).

A pessimistic variant of this is Robinson's (2012) argument that conservation NGOs should engage with businesses and markets because they are so dominant, even if the more likely impact of engagement is a reduced negative impact on the environment rather than a more positive conservation outcome. Other examples in the literature have shown that conservation NGOs sometimes conceive of farmers and other rural inhabitants as rational market actors, and therefore conservationists must engage with the rationality of market mechanisms (e.g., payments for ecosystem services) if they want to influence these people's behaviour and thus environmental impact (Wynne-Jones 2012). Yet, using market mechanisms to regulate people's behaviour does not just reflect that NGOs consider that these rural people frame their lives in terms of market exchanges and economic rationality, but also that the NGOs themselves are happy to frame their own intervention in these same terms.

In other examples, conservation NGOs' interventions using market mechanisms have the effect, intended or otherwise, of creating market subjects of the people affected (Dressler et al. 2013; Doyon and Sabinot 2014). In other words, by forcing market-based conservation on them, NGOs turned locals into something more closely resembling rational market actors than the people were prior to the intervention (see Fletcher 2010). In an argument somewhat resembling the Jujutsu argument, whale-watching enterprises and the NGOs who support them present it as pro-conservation capitalism displacing anticonservation capitalism in the form of whale hunting. By using markets to give economic value to whales as tourist resources, they can undermine the economic value of whales as a hunting resource, although this belies the economic complexities of why hunting declined, which was entirely incidental to the rise of whale tourism (Neves 2010). Whale-watching operators are eager to argue that whale watching is a means to bring about the end of hunting, by providing an alternative, and using this to shut down arguments about the ecological impact of whale watching. Both whale hunting and whale tourism turn whales into commodities and fetishise their ecological impact, hiding the negative effects.

The following section explores how conservation NGOs in southern Chile, particularly those associated with private PAs, have considered how and why they should engage with markets. It is based on detailed interviews with 47 key actors in NGOs and related organisations (e.g., government, ecotourism operators) operating in the region.

The section begins by outlining why Chile presents a useful case study for understanding the relationship between capitalism and conservation. It then explores the story of private PAs, particularly their emergence and diversity, focussing mainly on those owned and managed by national and international NGOs, as well as the ongoing attempts to create conservation easements in Chile. Through subsequently exploring the reasons for the emergence of these trends, the following section illustrates how the emergence and shape of Chile's private conservation initiatives reflects a kind of conservation Jujutsu, using the market to save nature from the market.

Why Chile?

Chile is an ideal place to begin exploring the relationship between markets and NGOs. During the military dictatorship of Augusto Pinochet (1973–1990), Chile saw a series of economic and political reforms with the aim of realigning the entire economy and society, initially according to Weberian principles, and then, under the guidance of Chicago School economists, along the lines of monetarist neoclassical economics. As a result, Chile is arguably the first truly neoliberal economy. These reforms have become deeply entrenched so that although ostensibly centre-left governments have been in power for much of the period since the return of democracy, there have only been extremely modest retreats from the principles and approaches of the Pinochet-era reforms (Valdés 1995; Murray 2002).

The environment and natural resources were particularly affected. The export of natural resources, mainly agricultural produce, copper and forestry products, were seen as the key to Chile's economic growth and were greatly incentivised. Key parts of the natural resource economy, such as forestry plantations, were greatly subsidised (Niklitschek 2007; Klubock 2014). Regulations were significantly reduced, and even in the democratic era, environmental impact assessments have been more about facilitating and approving development rather than controlling it (Tecklin et al. 2011; Latta and Aguayo 2012). Property rights were restructured to facilitate large-scale exporters at the expense of domestic small-scale producers.

State land holdings were privatised, individual land rights strengthened at the expense of collective land rights, and restrictions on foreigners holding land almost entirely abolished (Valdés 1995; Murray 2002). In a context where the state is unwilling to eagerly engage in conservation, and where natural resources are overwhelmingly viewed in monetary terms, there is a vacuum for both NGOs and private actors to become involved in conservation, and a situation in which market-based approaches may be treated by and large favourably.

Although many analyses consider that the process of neoliberalisation involves NGOs and other civil society actors taking roles previously allocated to the state, this is not the case in Chile. Here NGOs were suppressed during the military dictatorship, including environmental NGOs, and Chile emerged into the 1990s with a weaker and diluted conservation NGO sector compared to other countries (Carruthers 2001).

The Rise and Rise of Private Protected Areas in Chile

Private protected areas, those governed not by states or communities but by private actors, such as NGOs, corporations, individuals, families or cooperatives, began to emerge in southern Chile immediately after the return to democracy, although some were created under the dictatorship. The key actor was the United States (US) clothing entrepreneur Douglas Tompkins, who, after retiring from business, began purchasing tracts of land in southern Chile; in 1994 he declared the creation of Parque Pumalin. It is approximately 275,000 hectares and is one of the world's largest private PAs. The park is located at the northern edge of Chile's southern Patagonian area. Tompkins had had a long association with southern Chile and had conducted several mountaineering expeditions to the region in the 1960s. His declaration of the area as a park generated immediate controversy. It stretched from the Argentine border to the Pacific Ocean, effectively bisecting the country, raising questions about whether it was a threat to national sovereignty as well as a barrier to creating roads and other infrastructures between the southern and central regions[2] (Humes 2002; Holmes 2014).

Tensions over sovereignty were particularly acute given the long-running tensions with Argentina over the location of the border, and the political context of a country only recently emerging from military rule. Furthermore, although Tompkins saw his actions as like those of other wealthy entrepreneurs purchasing land for conservation in the US (see Jacoby 2001; Fortwangler 2007), such wild land philanthropy was unprecedented in Chile, so much so that his motivations were questioned by the public and by politicians. These concerns were exacerbated by the opaque manner by which Tompkins had purchased the land—that is, using a series of intermediaries ostensibly to avoid sellers raising their prices.

Among those who took his conservation intentions at face value, there were concerns that he was locking up Patagonia's natural resources, which should be used to further the development and expansion of the regional and national economy (Holmes 2014). This resulted in an agreement being signed between him and the then-president in which Tompkins promised to refrain from purchasing more land in the region and to allow nationally important infrastructures to cross Pumalin[3] (Nelson and Geisse 2001).

This agreement clashed significantly with the encouragement and freedom given to all other foreign buyers interested in purchasing Chile's natural resources, as part of the neoliberal economy and the focus on natural resources-led growth. The agreement was annulled shortly after its signing, and Tompkins subsequently purchased much more land in Chile and neighbouring Argentina, including slightly expanding Pumalin; however, this reflects that the political and public hostility to private PAs continues to this day in Chile.

Although, as explored later, this has abated substantially in recent years, such hostility has constrained the activities of private protected areas in Chile, particularly in their attempts to get government recognition and support (Holmes 2015). An illustration of this hostility is that, although the 1994 Base Law of the Environment—the key piece of post-dictatorship environmental legislation—includes a clause that the state should recognise and encourage private conservation efforts, no such recognition or incentives were forthcoming until proposals were put to Chile's parliament in 2014.

In subsequent years, more than 300 private PAs have been established in Chile, covering approximately 1,607,000 hectares, equivalent to 2.12% of Chile's land mass; remarkably, almost all these are located in southern regions (Holmes 2014, 2015). Foundations controlled by the Tompkins family created protected areas covering 45.5% of the area under private conservation. Large areas dominate—the 10 major private PAs constitute 41% of the total area conserved by private actors. Nonetheless, overall Chilean private protected areas could be defined by their heterogeneity (Tecklin and Sepulveda 2014). They vary greatly in size, from a few dozen hectares to more than 300,000 hectares. The PAs are held by a variety of owners such as individuals, universities (that use them as research stations), NGOs, cooperatives and businesses (Holmes 2015).

Many of the smaller areas owned by individuals or families have no business activities and are purely for conservation or family recreation purposes; others have limited commercial activities (usually tourism-related) to partially offset the cost of running the areas. Several larger PAs seek to make a profit. Some sites aim to combine tourism activities with real estate projects, creating limited developments set within a larger protected environment. Some are supplementing these with other income schemes, such as selling carbon credits on the voluntary market, based on reforestation of the properties with native trees. These enterprises emerged in the late 2000s and have been evolving and developing diverse income streams in an attempt to successfully combine conservation and capitalism (Holmes 2014, 2015).

A dominant theme has been real estate because land in southern Chile has been seen as a safe and profitable investment, increasing steadily in value during a time of global economic turmoil. One interesting example of this is the Huilo-Huilo Biological Reserve, a 60,000 hectare property dedicated to forestry; however, in the face of declining wood extraction, it has been transitioning into a private PA using tourism and limited real estate developments to generate income. In addition to companies for whom private protected areas are their main business, forestry enterprises own and manage some small PA properties as part of their sustainable forestry certification schemes and corporate responsibility programmes (Holmes 2015).

Several small Chilean NGOs also have established small private protected areas, mostly less than 1000 hectares in size, although one extends to 7500 hectares. The most significant property owned by a Chilean NGO is Parque Tantauco (118,000 hectares) in the Chiloé Archipelago. This property had been used for a series of failed forestry ventures; the Tantauco reserve was established in 2005 by Chilean airline and banking billionaire Sebastián Piñera through his NGO Fundación Futuro. He was an extremely well-known businessman who later served as Chilean president from 2010–2014. This property is significant in several ways. The park was the first very large private PA established by a Chilean national. It was explicitly inspired by the Tompkins model, and Doug Tompkins's wife attended the opening of Tantauco.

As such, the creation of Tantauco represented the changing political acceptability of private PAs, and the Tompkins family, among many Chileans and the Chilean political elite. Tantauco has had minimal commercial activity beyond a token entry fee and is funded almost entirely by Fundación Futuro, which is financed entirely by Piñera. There are indications that other wealthy Chileans may be purchasing large tracts of land in the far south for conservation, although details of these projects have yet to emerge (de la Fuente 2010).

There are two properties owned by large international NGOs, and both owe their origin to crises in the forestry sector. The first of these, the Valdivian Coastal Reserve (60,000 hectares), was established in 2005. Originally a property owned by a forestry company, the area and its surrounds were considered a priority area by The Nature Conservancy and other conservation NGOs because of its high levels of endemic biodiversity, and the threats posed by forestry activities and infrastructure developments, particularly proposed new roads. As the company that owned the property entered bankruptcy, a coalition of NGOs led by TNC, but with some financing and other assistance from both WWF and Conservation International (CI), formed to purchase the land at auction to create a protected area. Approximately 10% of the property is planted with exotic eucalyptus, and on maturation, this will be harvested and sold; the proceeds will be used to start a trust fund to finance the area. Through this, the intent is for the reserve to be self-financing (Holmes 2014).

To the far south, on the island of Tierra del Fuego, Karukinka National Park (275,000 hectares) was originally purchased by a US company, Trillium, with the intention of sustainably logging the native forests through careful long-rotation harvesting. The project proved too difficult and entered bankruptcy in 2002. Ownership passed to the creditor, the investment bank Goldman Sachs. Goldman Sachs initially entered discussions with TNC regarding the future of the site—Henry Paulson, Goldman Sachs's then chairman, was simultaneously the vice-chair of the Conservancy; however, at that point The Nature Conservancy was under significant criticism over its financial affairs following a series of investigative reports in the *Washington Post* (summarised in Stephenson and Chaves 2006).

Instead, Goldman Sachs donated the property to the Wildlife Conservation Society (WCS); Henry Paulson's son was a trustee of the WCS at the time. Karukinka has tried to self-finance its activities through a combination of a trust fund, seeded by donations by Goldman Sachs and a large personal donation by Henry Paulson, and through carbon credits from the substantial peat reserves on the property, although the latter plans have yet to materialise (Holmes 2015).

Recently, there have been some developments in attempts to get legal recognition for private PAs. In an example of 'policy entrepreneurship' (Hrabanski et al. 2013), The Nature Conservancy has led efforts to import a variant of conservation easements, which would allow landowners to voluntarily put permanent restrictions on what kinds of land use are permissible on their properties. A present vulnerability in private protected areas is that their conservation status depends on the willingness and ability of their current owners to conserve land; however, such easements would ensure that conservation efforts have a lifespan beyond current owners. Although a bill to create easements was introduced to Chile's parliament in 2014, the proposal is not a high political priority and has yet to pass into statute.

Some concern was raised by conservative politicians that such easements would lock up natural resources that should rightfully be used for economic growth. As a reaction to this, The Nature Conservancy and others promoting easements have been careful to cultivate cross-party support by arguing that easements and conservation are apolitical. As such, they can argue that easements do not challenge neoliberal arguments about economic growth or property rights.

There are multiple reasons for the emergence of private PAs. A key factor attracting Tompkins and some other investors, and a key factor in explaining why private conservation is disproportionately found in southern regions of Chile, is that the landscape aesthetics of snow-capped volcanoes, blue rivers and dense green forest are immensely attractive. Second, trends in land prices are very important. In southern regions until recently, prices were very low and properties very large, making it easy to purchase immense tracts of land relatively inexpensively; for example, it is rumoured that the purchase of Pumalin cost Tompkins some USD20 million. Thus far, prices have been increasing, rising in Patagonia by 115% between 2006 and 2011 (Jose Tapia and Muñoz 2012), making land purchases a sound investment for wealthy and middle-class individuals with an interest in conservation. The financial models of the for-profit private PAs, with their emphasis on real estate development, make use of this concept of property as a secure investment. There are several real estate agents who specialise in sales of large properties in Patagonia, for conservation and other purposes.

Third, private individual efforts to conserve are preferred because there is little confidence in other conservation routes, such as collective action or campaigns to tighten environmental regulation, although this is beginning to change because of high-profile environmental crises (Sepúlveda and Villarroel 2012). Such regulations in Chile during the dictatorship and the first two decades of democratic rule have been characterised as market-enabling, facilitating the exploitation of natural resources for businesses and large infrastructure projects rather than focussing on regulating negative impacts or preventing harmful developments (Tecklin et al. 2011).

A consequence of this structure, which lacked strong central regulations, was that environmental protection devolved into individual environmental impact assessments, meaning that environemntalists fought a series of local battles over the assessment of individual projects rather than fighting national issues (Sepúlveda and Villaroel 2012). In addition, government policy on the environment is set by a narrow group of bureaucrats, politicians and business representatives that is difficult for outsiders to penetrate (Tecklin et al. 2011). As such, environmental campaigns have found it difficult to access key decision-making arenas and change discourses and policies.

One exception was the campaign to create conservation easements—this only had its success because those behind it could use their personal connections with senior figures in corporate law firms and chambers of commerce. The situation may be changing, particularly as high-profile campaigns and scandals have led to public pressure on the failures of environmental regulations and the impact of large infrastructure projects (Sepúlveda and Villaroel 2012). Finally, and perhaps most important, even as there may be barriers to collective action or focussing on regulation, there are also other measures that make individual action, particularly purchasing land, much easier in Chile than in other countries. The neoliberal reforms of the dictatorship, and their continuation in the democratic era, have created a situation where private property rights are sacrosanct, with very low levels of corruption, as well as minimal restrictions on foreign land ownership. These measures, introduced to allow national and foreign buyers to purchase land to extract natural resources, also have allowed the purchase of land to conserve natural resources.

The breadth of private conservation initiatives in Chile reflect a broad range of views on the merits of using markets to make conservation pay for itself. Some, such as Tompkins and their NGOs, are staunchly against such an approach—the operations director of Pumalin commented in an interview that 'I think that the goal of self-financing is absurd, because in order to self-finance a national park it would have to be Disney world. There isn't a park in the world that'. Nevertheless, others see markets, capitalism and profit as complementary—one business operating a private PA has as its slogan 'for profit conservationists'. All of them, however, are buying nature on the open market to save it from other market actors, even if not all of them are subsequently selling nature to save it (see McAfee 1999).

There are no direct or indirect subsidies or other measures that give conservationists an advantage compared to other actors when purchasing property—they must merely be the highest bidder. The Valdivian Coastal Reserve, Tantauco, and several less important properties owned by smaller NGOs and less-wealthy individuals were purchased to prevent them from being acquired by forestry companies, either preemptively or in response to forestry firms' bankruptcies and crises. In the case of Karukinka, Goldman Sachs came under pressure from some shareholders and other actors to sell the land to another forestry company rather than to donate it for conservation. Those NGOs and other actors that purchase land for conservation in southern Chile are practising conservation Jujutsu by using market structures intended to facilitate the exploitation of natural resources to instead protect and conserve them.

The strength of NGOs and conservationists should not be overstated here. Their Jujutsu is a reaction to the political economy in which they are operating, an attempt to exploit it for their purposes, rather than dominating and creating these same circumstances in the way that businesses can. They have fewer political connections; have fewer financial resources; and own much less land than the forestry, hydroelectricity and fisheries sectors that exploit natural resources (Holmes 2014). They must conform, at least in public, to the dominant discourses of economic growth based on exploitation of natural resources as part of their bid to get public legitimacy and political support for campaigns such as the introduction of easements. For information on conservationists engaging with markets to broaden the constituency for conservation, see Fisher and Brown (2014).

Conservationists note that the political class believe that economic growth is paramount and that conservation is antigrowth. One prominent owner of a private PA noted in an interview that 'they [the political class] very much believe in the economy drives everything, and conservation is, … they haven't got into this new paradigm where conservation is a necessary part of growth. They still believe that conservation is against economic growth, you either conserve or you grow'. Thus, rather than campaigning on the intrinsic value of conservation, politicians feel that they have to emphasise its economic value. Another prominent individual in the private protected area sector in an interview noted:

> [When] you speak to politicians, who generally assess things as economists, you have to compete directly with these values. So, a forestry company will say, 'listen, I can support GDP with so many millions of dollars, or the local economy with so many millions of dollars', and the conservationists say 'I support three little frogs by conserving them'. So, sadly, you need to enter this logic of saying 'I conserve water worth so many millions of dollars.

Conservationists are embracing markets (through private land purchasing) as the best way to do conservation, embracing the Jujutsu approach as the most feasible means of conserving biodiversity, given the circumstances, although only some of them have enthusiastically grasped markets and established a significant amount of market-based activities (e.g., through ecosystem services, ecotourism, controlled real estate or land speculation). As such, these could be considered as advancing the commodification of natural resources; extracting profit from nature; and incorporating it in global circulation of capital, as does forestry, fisheries and other primary industries. Nevertheless, the majority of private protected areas do not seek to make a profit and instead are intending to slow down or limit the capitalist exploitation of natural resources, even though their success in doing so may be rather limited (Holmes 2015). Buying land for conservation does not fundamentally undermine the causes of natural resource degradation, mainly attributed to the expansion of lightly regulated primary industries, but it does limit their reach in certain places.

Furthermore, private PAs in Chile can only offer more than other land uses where they are able to do so, itself a reflection of the demand for the land from natural resource industries. Thus, private protected areas may be located in places of relatively low threat. Tantauco and Karukinka were places where forestry enterprises had struggled, perhaps because they are less than optimal places for the prevailing forms of forestry (Tecklin and Sepulveda 2014). Pumalin is located on marginal land, with steep inclines and remote terrain, limiting the possibilities for forestry.

Several owners of private PAs stated that they purchased the land for a relatively low price because the previous owners struggled to make money from farming on the sites. There are very few private protected areas in the central regions of Chile, where demand from the agricultural sector keeps land values high. As such, private PAs could be considered to be providing a limited challenge both to the incorporation of Chile's natural resources into global circulation of capital and to the loss of its biodiversity.

Concluding Remarks

We should expect a level of complexity when exploring conservationists' engagement with markets. Conservation is not homogenous and individual NGOs are not homogenous. A variety of views for why conservationists should engage with markets, and why markets should engage with capitalism are to be expected. There are two main explanations for the rise of neoliberal conservation: (1) as capitalism looks to nature as a new frontier for generating surplus and (2) as conservationists turn to market-based tools as the most effective means to save biodiversity in a market-dominated world. In Chile, both are visible: conservation (through ecotourism and carbon trading) is just another way of making money from the natural resources of southern Chile, alongside forestry and fisheries, while conservationists are using markets, particularly on private land, as a pragmatic way of saving nature in a strongly neoliberal country. The latter is more visible than the former, but both have been facilitated by the reforms of the Pinochet era (Holmes 2015).

At one end of this pragmatic view lies the Jujutsu position, which seeks to use markets (private land purchasing) to gain control over natural resources and conserve them either by reselling in conservation-friendly forms, such as payments for ecosystem services (e.g., Karukinka), or going further and removing them from additional involvement in global capitalist circulation (e.g., Pumalin).

It is possible to imagine that the Jujutsu approach of pragmatically engaging with markets to undermine conservation-unfriendly markets in natural resources could be a viable conservation approach when land is economical, leaving an opportunity for conservationists to step in. Southern Chile has been a resource frontier in recent decades, rapidly driving up land prices and limiting the opportunities for conservationists, making Jujutsu progressively more difficult with time. There is some evidence that private interventions have been more effective than state interventions, with private PAs more likely to protect endangered habitat than state protected areas (Pliscoff and Fuentes-Castillo 2011). Nonetheless, private protection may not be providing significant benefits to humans.

Many, particularly the Tompkins properties, are focussed on wilderness protection, with little evidence that there is potential for something directed toward immediate human benefits. Indeed, there are several conflicts between local communities and private PAs, although it is unclear whether these are more prevalent than conflicts with state protected areas, and they are certainly rarer than conflicts with natural resource extractors such as forestry companies (Holmes 2014).

The large, powerful big international NGOs (BINGOs) are involved only in private PAs in Chile by accident, not necessarily as part of a wider global strategy. Karukinka became a private area owned by the WCS only as a condition of a donation, not necessarily because it was the Society's preferred option. The Valdivian Coastal Reserve was a reactive strategy; while the WWF had long sought to conserve the area, it only created a PA as a reaction to a bankruptcy, not as a planned solution. Indeed, the WWF's constitution prevents it from owning large amounts of land for conservation. Few NGOs globally engage in private land conservation because it is expensive, difficult and controversial (Holmes 2013). Instead, these large NGOs are much more enthusiastic, at least at the leadership level, about neoliberal forms of conservation and the potential for solutions, which work for capitalism and for nature, than the Jujutsu approach to undermine markets in nature (Brockington and Duffy 2010).

Despite this, it is worth using the concept of Jujutsu conservation when analysing trends in neoliberal conservation. Doing so makes it possible to analyse the rationale and tactics used by conservation organisations in detail as they engage with markets, rather than just relying on broad-brush theoretical approaches to capitalism as a way of explaining the roots of the trends observed.

Notes

1. I am indebted to Mike Mascia for the comparison between markets in conservation and Jujutsu.
2. Even to this day, there are no continuous roads between sparsely populated Chilean Patagonia and the central provinces where the majority of the population reside. Travellers must go by sea, air or travel by road via Argentina. The poor infrastructure is because the low and dispersed population in the southern region makes such an infrastructure a poor investment, and because the steep and seismically active terrain provides an extreme engineering challenge. Parque Pumalin is one of several gaps in connectivity.
3. Parque Pumalin is in two discontinuous segments. The property bisecting Pumalin was originally owned by a Chilean university (Pontificate University of Valparaiso) that were put under political pressure not to sell to Tompkins, and a Chilean buyer was sought, particularly among utility companies that might want to build an electricity infrastructure to connect the southern regions, with their immense hydroelectricity potential, to the central regions and power-hungry mines of the north. In an ironic development, ownership of the property was transferred to a foundation controlled by the university and by a Chilean utility company, later bought out by a Spanish-Italian conglomerate, and the area was declared as a private protected area. The property, like Pumalin, is a private PA owned by a foundation controlled by foreigners, precisely the situation that the presidential intervention was intended to prevent.

References

Blanchard, L., C.G. Sandbrook, J.A. Fisher, and B. Vira. 2016. Investigating the Consistency of a Pro-market Perspective Amongst Conservation Professionals: Evidence from Two Q Methodological Studies. *Conservation and Society* 14 (2): 112–124.

Brockington, D., and R. Duffy. 2010. Capitalism and Conservation: The Production and Reproduction of Biodiversity Conservation. *Antipode* 42 (3): 469–484.

Brockington, D., and K. Scholfield. 2010. Expenditure by Conservation Nongovernmental Organizations in Sub-Saharan Africa. *Conservation Letters* 3 (1): 106–113.

Brosius, J.P., and L.M. Campbell. 2011. Collaborative Event Ethnography: Conservation and Development Trade-offs at the Fourth World Conservation Congress. *Conservation and Society* 8 (4): 245–255.

Büscher, B., S. Sullivan, K. Neves, J. Igoe, and D. Brockington. 2012. Towards a Synthesized Critique of Neoliberal Biodiversity Conservation. *Capitalism Nature Socialism* 23 (2): 4–30.

Carruthers, D. 2001. Environmental Politics in Chile: Legacies of Dictatorship and Democracy. *Third World Quarterly* 22 (3): 343–358.

Castree, N. 2008a. Neoliberalising Nature: Processes, Effects, and Evaluations. *Environment & Planning A* 40: 153–173.

———. 2008b. Neoliberalising Nature: The Logics of Deregulation and Reregulation. *Environment & Planning A* 40: 131–152.

Corson, C. 2010. Shifting Environmental Governance in a Neoliberal World: US AID for Conservation. *Antipode* 42 (3): 576–602.

de la Fuente, A. 2010. El nuevo mapa de la conservación. *Que pasa* 11 June.

Doyon, S., and C. Sabinot. 2014. A New 'Conservation Space'? Protected Areas, Environmental Economic Activities and Discourses in Two Yucatán Biosphere Reserves in Mexico. *Conservation and Society* 12 (2): 133.

Dressler, W.H., P.X. To, and S. Mahanty. 2013. How Biodiversity Conservation Policy Accelerates Agrarian Differentiation: The Account of an Upland Village in Vietnam. *Conservation and Society* 11 (2): 130.

Fisher, J.A., and K. Brown. 2014. Ecosystem Services Concepts and Approaches in Conservation: Just a Rhetorical Tool? *Ecological Economics* 108: 257–265.

Fletcher, R. 2010. Neoliberal Environmentality: Towards a Postructuralist Political Ecology of the Conservation Debate. *Conservation and Society* 8 (3): 71–181.

Fortwangler, C. 2007. Friends with Money: Private Support for a National Park in the US Virgin Islands. *Conservation and Society* 5 (4): 504–533.

Holmes, G. 2010. The Rich, the Powerful and the Endangered: Conservation Elites, Networks and the Dominican Republic. *Antipode* 42 (3): 624–646.

———. 2011. Conservation's Friends in High Places: Neoliberalism, Networks, and the Transnational Conservation Elite. *Global Environmental Politics* 11 (4): 1–21.

———. 2013. *What Role Do Private Protected Areas Have in Conserving Global Biodiversity?*. SRI working papers (46).

———. 2012. Biodiversity for Billionaires: Capitalism, Conservation and the Role of Philanthropy in Saving/Selling Nature. *Development and Change* 43 (1): 185–203.

———. 2014. What Is a Land Grab? Exploring Green Grabs, Conservation, and Private Protected Areas in Southern Chile. *Journal of Peasant Studies* 41 (4): 547–567.

———. 2015. Markets, Nature, Neoliberalism, and Conservation Through Private Protected Areas in Southern Chile. *Environment & Planning A* 47 (4): 850–866.

Hrabanski, M., C. Bidaud, J.-F. Le Coq, and P. Méral. 2013. Environmental NGOs, Policy Entrepreneurs of Market-Based Instruments for Ecosystem Services? A Comparison of Costa Rica, Madagascar and France. *Forest Policy and Economics* 37: 124–132.

Humes, E. 2002. *Eco Barons: The Dreamers, Schemers, and Millionaires Who Are Saving Our Planet*. New York: Ecco/HarperCollins.
Igoe, J., and D. Brockington. 2007. Neoliberal Conservation: A Brief Introduction. *Conservation and Society* 5 (4): 432–449.
Igoe, J., and C. Fortwangler. 2007. Whither Communities and Conservation? *International Journal of Biodiversity Science & Management* 3 (2): 65–76.
Igoe, J., K. Neves, and D. Brockington. 2010. A Spectacular Eco-Tour Around the Historic Bloc: Theorising the Convergence of Biodiversity Conservation and Capitalist Expansion. *Antipode* 42 (3): 486–512.
Jacoby, K. 2001. *Crimes against Nature: Squatters, Poachers, Thieves, and the Hidden History of American Conservation*. Berkeley: University of California Press.
Jose Tapia, M. and M. Muñoz. 2012. Tierras en Aysen duplican en valor por conservacionismo y proyectos electricos. *La Tercera*. 31 March.
Kareiva, P. 2014. New Conservation: Setting the Record Straight and Finding Common Ground. *Conservation Biology* 28 (3): 634–636.
Kareiva, P., C. Groves, and M. Marvier. 2014. Review: The Evolving Linkage Between Conservation Science and Practice at The Nature Conservancy. *Journal of Applied Ecology* 51 (5): 1137–1147.
Klubock, T. 2014. *La Frontera: Forests and Ecological Conflict in Chile's Frontier Territory*. Durham: Duke University Press.
Lalasz, R., P. Kareiva, and M. Marvier. 2011. Conservation in the Anthropocene. *Breakthrough Journal* 2: 26–36.
Latta, A., and B.E.C. Aguayo. 2012. Testing the Limits Neoliberal Ecologies from Pinochet to Bachelet. *Latin American Perspectives* 39 (4): 163–180.
MacDonald, K. 2010. The Devil Is in the (Bio)diversity: Private Sector 'Engagement' and the Restructuring of Biodiversity Conservation. *Antipode* 42 (3): 513–550.
Mbembé, J.-A. 2001. *On the Postcolony*. Stanford: University of California Press.
McAfee, K. 1999. Selling Nature to Save It? Biodiversity and Green Developmentalism. *Environment and Planning D: Society and Space* 17 (2): 133–154.
Murat, A., and B. Büscher. 2012. Nature™ Inc.: Changes and Continuities in Neoliberal Conservation and Market-based Environmental Policy. *Development and Change* 43 (1): 53–78.
Murray, W. 2002. From Dependency to Reform and Back Again: The Chilean Peasantry During the Twentieth Century. *The Journal of Peasant Studies* 29 (3–4): 190–227.
Nelson, M., and G. Geisse. 2001. Las lecciones del caso Tompkins para la política ambiental y la inversión extranjera en Chile. *Ambiente y desarrollo* 17 (3): 14–26.
Neves, K. 2010. Cashing in on Cetourism: A Critical Ecological Engagement with Dominant E-NGO Discourses on Whaling, Cetacean Conservation, and Whale Watching. *Antipode* 42 (3): 719–741.

Niklitschek, M.E. 2007. Trade Liberalization and Land Use Changes: Explaining the Expansion of Afforested Land in Chile. *Forest Science* 53 (3): 385–394.

Peck, J., and A. Tickell. 2002. Neoliberalizing Space. *Antipode* 34 (3): 380–404.

Pliscoff, P., and T. Fuentes-Castillo. 2011. Representativeness of Terrestrial Ecosy stems in Chile's Protected Area System. *Environmental Conservation* 38: 303–311.

Robinson, J.G. 2012. Common and Conflicting Interests in the Engagements Between Conservation Organizations and Corporations. *Conservation Biology* 26 (6): 967–977.

Sandbrook, C., I.R. Scales, B. Vira, and W.M. Adams. 2011. Value Plurality Among Conservation Professionals. *Conservation Biology* 25 (2): 285–294.

Sepúlveda, C., and P. Villarroel. 2012. Swans, Conflicts, and Resonance: Local Movements and the Reform of Chilean Environmental Institutions. *Latin American Perspectives* 39 (4): 181–200.

Soule, M. 2013. The 'New Conservation'. *Conservation Biology* 27 (5): 895–897.

Stephenson, M., and E. Chaves. 2006. The Nature Conservancy, the Press, and Accountability. *Nonprofit and Voluntary Sector Quarterly* 35 (3): 345–366.

Tecklin, D.R., and C. Sepulveda. 2014. The Diverse Properties of Private Land Conservation in Chile: Growth and Barriers to Private Protected Areas in a Market-friendly Context. *Conservation and Society* 12 (2): 203–217.

Tecklin, D., C. Bauer, and M. Prieto. 2011. Making Environmental Law for the Market: The Emergence, Character, and Implications of Chile's Environmental Regime. *Environmental Politics* 20 (6): 879–898.

Valdés, J.G. 1995. *Pinochet's Economists: The Chicago School of Economics in Chile*. Cambridge, UK: Cambridge University Press.

Wynne-Jones, S. 2012. Negotiating Neoliberalism: Conservationists' Role in the Development of Payments for Ecosystem Services. *Geoforum* 43 (6): 1035–1044.

CHAPTER 8

Strengths and Limitations of Conservation NGOs in Meeting Local Needs

Emmanuel O. Nuesiri

Introduction

Nongovernmental organisations (NGOs) have become key actors in the design and implementation of environmental conservation initiatives around the world. The NGOs' representatives can be found seated at the table at international conferences deciding on the fate of tropical forests, at national-level dialogues designing national environmental action plans, and at local levels implementing community-based conservation initiatives. Although there is no agreed-on definition of 'NGOs', they are neither state nor private sector institutions, but work for the public good in environmental conservation and/or the social development sector (Brockington and Scholfield 2010). The NGO world includes international giants, such as the World Wide Fund for Nature (WWF), which enjoys diplomatic status in a number of countries, and the International Union for Conservation of Nature (IUCN) with a membership that includes national governments. However, the NGO world also includes countless small organisations with limited budgets working in local communities in the developing world (Brockington and Scholfield 2010).

E.O. Nuesiri (✉)
University of Potsdam, Potsdam, Germany

P.B. Larsen, D. Brockington (eds.), *The Anthropology of Conservation NGOs*, Palgrave Studies in Anthropology of Sustainability, DOI 10.1007/978-3-319-60579-1_8

International NGOs often partner with multilateral, bilateral and private donors including the World Bank, United Nations agencies (e.g., the UN Development Programme), and national country governments to implement their programmes. These NGOs also work as service providers for donor institutions, international organisations and governments that recruit them to assist in designing and/or implementing environmental policy. Likewise, smaller, local NGOs work to execute their own initiatives and projects, often with donor funding; and they also work as service providers recruited by larger NGOs, donors, international organisations and governments to assist with service delivery at the local level. Both conservation and development sector NGOs have become adept at local service delivery, so much so that some have opined that they could even be recruited to replace local government authorities as the substantive representatives of local people (UNDDSMS and UNDP 1996).

This chapter thus examines the strength and limitations of NGOs as the political representatives of local people. Although it acknowledges that NGOs are broadly divided into development sector NGOs and conservation sector NGOs (Brockington and Scholfield 2010), it is based on a study of conservation sector NGOs. It presents an analysis of the role of conservation NGOs in local environment and development initiatives informed by the theory of political representation (Pitkin 1967; Manin et al. 1999; Mansbridge 1999; Saward 2006, 2008; Urbinati and Warren 2008; Rehfeld 2011; Montanaro 2012). The chapter proceeds through a case study of the involvement of Nigerian conservation NGOs in the country's programme on reducing emissions from deforestation and forest degradation with the added goals of conserving and enhancing forest carbon stocks and sustainably managing forests—Reducing Emissions from Deforestation and Forest Degradation (REDD+, hereafter Nigeria-REDD).

Nigeria is a federal republic with 36 states and a government based in Abuja, its federal capital territory. The country's REDD+ Programme has its origin in the June 2008 Cross River State Stakeholders Summit on the Environment hosted by the state's governor, Senator Liyel Imoke, in Calabar, the Cross River State capital (CRS 2008). Cross River State (CRS) includes about 50% of Nigeria's primary tropical forest and is the home of the Cross River National Park (CRNP), the largest tropical forest protected area (PA) in Nigeria (Oyebo et al. 2010). The Cross River forest is also home to the critically endangered Cross River Gorilla (*Gorilla gorilla diehli*) and other endangered primates of global conservation value

(Nuesiri 2003; Oates et al. 2004, 2007). At the CRS Summit, influential local actors led by the Calabar-based NGO Coalition for the Environment (NGOCE) recommended that the governor ban commercial logging in the state and opt for carbon forestry (Nuesiri 2015).

Governor Imoke accepted the carbon forestry recommendation and proceeded to ban commercial logging in the CRS; in October 2009 he attended a REDD+ capacity-building workshop hosted by the Katoomba Group (an international NGO promoting carbon forestry) in Ghana. In November 2009 he requested that the Nigerian Federal Ministry of Environment (FME) join the UN-REDD, and in December 2009 he attended the Fifteenth Conference of the Parties (COP 15) in Copenhagen where he requested support for Nigeria to design a REDD+ readiness programme. In March 2010, the UN-REDD approved Nigeria's membership and in October 2011, at its Seventh Policy Board Meeting, granted Nigeria USD4 million to design a REDD+ readiness programme. Nigeria-REDD therefore has a state-level programme with the Cross River State as the pilot location for forestry activities and a national-level programme for policymakers (FME 2012; Nuesiri 2015).

Research methods for this study included 'grey' and published literature review and semistructured interviews held during field work in the summer of 2012 and 2013 in Nigeria. The researcher interviewed a total of 125 participants drawn from local communities, Nigeria-REDD, UN-REDD and local NGO staff. Research methods also included three focus group meetings with personnel from three NGOs (one local, one national, one international), and participant observation while attending a meeting of the Akamkpa Council of Chiefs to which this researcher was invited for a question-and-answer session about REDD.

This chapter is organised as follows: 'Political Representation' is a review of political representation theory. It argues that NGOs fall into the category of symbolic political representatives able to speak for local people in decision-making spaces based on a shared subscription to social development discourses. The section, 'Conservation NGOs and Launching of the Nigeria-REDD Programme', examines the role of conservation NGOs during the design of the Nigeria-REDD readiness proposal, showing how they were manipulated into legitimising the consultative process that accompanied the proposal design. The next section, 'Limitations of NGOs Involved in Launching of Nigeria-REDD', discusses the limitations of NGOs involved in design of the Nigeria-REDD proposal, while the Concluding Remarks section summarises this chapter.

Political Representation

In nation-states today, policy decisions typically are made by the executive and legislative branches occupied by persons with claims to representing either single or grouped political constituencies. This type of representation is at the heart of political representation theory, which is defined as the act of political representatives being present in decision-making arenas, representing the interests of groups physically absent from that arena (Pitkin 1967; Runciman 2007). Political representation is democratic when there are accountability mechanisms, such as periodic elections, which ensures that representatives are responsive to the represented (Manin et al. 1999; Rehfeld 2006). It may be considered undemocratic when electoral choice, downward accountability and responsiveness to the development needs of the governed are absent. In cases when undemocratic regimes are responsive to the governed, they have been categorised as 'benign dictatorships' (Wintrobe 1998; Ribot et al. 2010) or 'good despotism' (Mill 1861: 48).

Pitkin (1967) discusses four types of political representation: (1) symbolic, (2) formal, (3) descriptive and (4) substantive. Symbolic representation is when an agent (i.e., object, institution or person) speaks for a group; constituency; or state based on a shared discursive, affective and aspirational relationship between the agent and the represented (Dryzek and Niemeyer 2008; Lombardo and Meier 2014). In this way, the national flag (an object) represents a nation, the NGO (an institution) speaks for and about the development needs of a local community, and a monarch or chief (a person) speaks for the subjects. The relationship between agent and represented is legitimised through discourses that appeal to both parties; for the flag, it is discourses of statehood; for NGOs, it is the development discourses that define their mission and resonate with the aspirations of local communities; for the monarch or chief, it is often discourses of identity. Thus, institutions (e.g., NGOs) can represent and speak for local communities but are not the principal actor statutorily responsible for social development at the local level; this mandate is meant for local government authorities.

Formal representation is when a state officially authorises an agent to represent a constituency to the state or to represent the state to an audience—for example, when national REDD+ focal points represent their governments in UN-REDD board meetings. Descriptive representation is when representatives are chosen based on their resemblance to the group they are standing for (Pitkin 1967); they are selected because they are considered 'typical of the larger class of persons whom they represent' (Mansbridge 1999: 629).

Substantive representation is when representatives act for and are accountable to the represented for their actions (Pitkin 1967). In this context, the represented have objective criteria by which to evaluate and sanction their representatives appropriately (Pitkin 1967). Substantive representation is considered morally superior to descriptive and symbolic representation for the checks it places on the actions of representatives (Pitkin 1967). Substantive representation is viewed as meeting the requirements of social justice, making it the preferred operational mechanism behind representative democracy (Mill 1861; Grunebaum 1981; Kateb 1981; Manin et al. 1999; Fraser 2005; Urbinati and Warren 2008; Rehfeld 2011). As members of government, substantive representatives are responsible for social and economic policies that are just and equitable and responsive to the social development needs of all groups in society. Elected local governments are the substantive representatives of local communities that are statutorily mandated to deliver social development in an accountable and responsive manner.

Representation theorists, such as Saward (2006, 2008), however, argue that representation is about claims-making—that elected and unelected representatives make claims about themselves, their constituency, and their relationship to their constituency, and that claims made by elected and unelected representatives have equal legitimacy. This is true insofar as the task of a representative is viewed as standing for and speaking for another, as is the case with descriptive and symbolic representation. Substantive representation goes beyond standing and speaking for others because there is a responsiveness–accountability relationship between the representative and the represented.

The responsiveness criteria are valid only in a context where the representative has access to resources that can be drawn on to meet the needs of the represented (Manin et al. 1999). Even though NGOs have access to grants and funds from service delivery contracts with all the uncertainties that accompany these types of funding, local government authorities have statutory regular access to public finance for local development purposes. Thus, sustainable long-term local social development cannot be built on an NGO institutional platform.

The NGOs are not just symbolic representatives but often subscribe to symbolic politics in carrying out their representative function. Symbolic NGO action involves the use of images, objects, theatrics and discourses that appeal to the emotions to influence decisions of the general public, private corporations, governments and international organisations (Keck and Sikkink 1999; Silveira 2001; Miller 2012). Whereas Edelman (1985) in his seminal work on symbolic politics argues that it is a mechanism for manipulating the public, Brysk (1995) shows that it is an effective mechanism employed by civil society organisations in their work. Bluhdorn (2007) provides a symbolic politics continuum showing the logic behind symbolic politics.

The continuum shown in Table 8.1 has 2 categories, 4 variables and 8 criteria for evaluating symbolic politics. The political action of the NGO spans all 8 criteria on the continuum. Bluhdorn's schema (2007) opens the black box that is symbolic politics, showing that the motives and manifestations are varied. This is a counterpoint to the influential work of Edelman (1985) who maintained that symbolic politics is subscribed to by governments to deceptively manipulate the public. Bluhdorn's work (2007) is more nuanced, making it a suitable empirical device for getting at the 'why' of symbolic actions from government, NGOs and customary authorities. While representation theory enables an understanding of the 'what' with regard to NGO representation, the Bluhdorn schema (2007) enables an understanding of the 'why'.

As symbolic representatives of local communities, most visibly when they advocate for governments to respond to development needs of local communities, NGOs have no mandate from the people to deliver social development. This has not stopped NGOs from trying when they can find the funding for projects; more often such are short-term projects with a life cycle rarely extending beyond five years. In this context, NGOs' strength lies in their ability to quickly adopt and demonstrate new approaches to meeting development needs. Keeping with their symbolic function in the political landscape, NGO projects serve as signage showing how to innovatively meet local needs for governments to adopt on a more long-term basis. The next section examines the role played by local NGOs in instituting a REDD+ readiness programme in Nigeria; the case study sheds more light on the strength and limitations of NGOs in meeting local needs.

Table 8.1 The symbolic politics continuum

Symbolic politics							
Political use of symbols (A)				*Replacement action (B)*			
Integration, mobilisation (A1)		*Complexity reduction (A2)*		*Best possible option (B1)*		*Deliberate strategy (B2)*	
Propagandistic instrumentalisation for elite agenda	For emancipatory project	In response to media needs	In response to citizen needs	In light of fragmented and contradictory citizen interests	In light of imperatives emerging from the political system	To temporarily delay the uncomfortable consequences of effective action	To deceive the public and pursue secret elite agendas
(A1.1)	(A1.2)	(A2.1)	(A2.2)	(B1.1)	(B1.2)	(B2.1)	(B2.2)

Source: Adapted from Bluhdorn (2007)

Conservation NGOs and Launching of the Nigeria-REDD Programme

Here we provide (1) a genesis of Nigeria-REDD and (2) the role of conservation NGOs in validating the Nigeria-REDD proposal. Nigeria has a federal system of government (FGN 1999), but its 36 states and 774 local government areas depend on financial allocations from the federal government for their operations (ARD 2001; Adesopo 2011; Omar 2012). Even though state government allocations are deposited directly into the state government bank account, local government allocations from the federal government get routed through the state government, giving state governors great influence over local governments (ARD 2001; Diejomaoh and Eboh 2010). The governors 'decide how the federal money gets used and … decides who gets the contracts and when to pay. … [I]t takes away the power of the local government chairmen' (Local bureaucrat interviewed in Calabar in 2012). Thus, governors maintain significant control over local governance arrangements in Nigeria.

In addition, the Nigeria Land Use Act of 1978 (FGN 1978) places land under the authority of state governors. Local governments cannot allocate land to a user of greater than 5000 hectares (50 km^2) without the consent of the governor; however, the governor does not need to consult the local government if it needs rural land for public purposes (FGN 1978). Nevertheless, unlike local governments, customary authority still exercises de facto rights over land use and land sales in rural communities. Thus, state governors often consult with local chiefs before enclosing public lands in rural communities, although state governors are not statutorily obliged to do so (USAID 2010). Land tenure arrangements therefore explain the control state governors have over land and consequently over forestry initiatives (e.g., REDD+). This also has important implications for the way local NGOs relate with local authorities in Nigeria.

Genesis of the Nigeria-REDD Programme

As mentioned earlier, the Nigeria-REDD Programme gets its roots from the June 2008 Cross River State Stakeholders Summit on the Environment hosted by the governor, Senator Liyel Imoke (CRS 2008). The Summit was used to assess how CRS forest resources could better contribute to generating revenue for the state. At that time, the state was under pressure to seek new sources of revenue because it had lost most of its crude oil concessions when Nigeria ceded the Bakassi peninsula to Cameroon in 2008 (Konings 2011). It also was locked in a legal dispute, which it eventually lost, with the neighbouring Akwa Ibom State over ownership of productive oil wells that fell into Akwa Ibom territory when it was carved out of the Cross River State (AKSG 2012). The loss of its petroleum resources translated into a loss of between 20% and 25% of its financial allocations from the federal government that averaged USD115 million[1] per year for the period 1999–2008 (Olubusoye and Oyedotun 2012).

The environmental Summit recommended that the Cross River State put a stop to commercial logging and instead develop a carbon forestry programme as an alternative biodiversity-friendly means to generate revenue (CRS 2008; Oyebo et al. 2010). This idea was championed by the Calabar-based NGO Coalition for the Environment (NGOCE), a well-known local NGO in the international forest conservation community. The idea was supported by the Nigerian Conservation Foundation (NCF), the largest Nigerian NGO; the Tropical Forest Group (TFG), an international NGO with a long-time presence in CRS; and by Environmental Resources Management (ERM), an international environmental consultancy with its head office in London (Filou 2011; TFG 2011; FME 2012).

The governor accepted the recommendations from the Summit, banned commercial logging in the state shortly after, and began working to design a forest carbon programme (Oyebo et al. 2010). The governor restructured the state Forestry Commission (the CRSFC), appointing an outsider from the NGO sector, Odigha Odigha, as the chairperson in 2009 (Filou 2010). Odigha Odigha won the Goldman Environmental Prize in 2003 as coordinator of NGOCE working against uncontrolled commercial logging in the Cross River State (Filou 2010).

Acting on his decision to develop a carbon forestry programme, Governor Imoke led a delegation from CRS, including members of the Forestry Commission, to the October 2009 Katoomba XV Meeting, 'Integrated Solutions: Water, Biodiversity and Terrestrial Carbon in West Africa' in Ghana (Oyebo et al. 2010). Katoomba is a working group set up by Forest Trends, a global NGO working to promote the adoption of payment for environmental services schemes, such as REDD, around the world (Forest Trends et al. 2008). The governor shared his vision for a REDD initiative at Katoomba XV and invited expert consultants from the meeting to come and work with the state Forestry Commission to draft a REDD project idea note (REDD PIN) for the Cross River State (Oyebo et al. 2010).

In November 2009, Governor Imoke requested that the Nigerian Ministry of Environment include REDD in its climate change mitigation strategy document and apply to the UN-REDD for membership. In December 2009, the governor and a delegation from the Forestry Commission attended the UNFCCC Conference of Parties meeting in Copenhagen (i.e., COP 15). At the meeting the governor gave a presentation in which he requested financial support to develop a REDD program in Nigeria (Oyebo et al. 2010). In January 2010, experts from Katoomba visited the REDD pilot sites in CRS to obtain data to produce the REDD Project Idea Note (R-PIN) for Nigeria (FME 2012).

By March 2010, Nigeria's membership request to the UN-REDD was approved and in October 2010, a UN-REDD mission visited Nigeria to contribute to the preparation of a Nigeria-REDD readiness proposal for submission to the UN-REDD Policy Board (FME 2012). On 18 February 2011, a draft REDD readiness proposal was presented to a participatory stakeholders' forum in Calabar chaired by the governor of Cross River State. The forum was attended by about 100 people, including a delegation from the UN-REDD and local NGOs in Calabar. A similar meeting was held on 21 February 2011 in Abuja with national-level actors. The document was then submitted to the UN-REDD on 22 February 2011 for consideration at its Sixth Policy Board Meeting in March 2011 (FME 2012).

The UN-REDD Policy Board requested changes in the document (Global Witness 2011; FME 2012), which were effected and a second participatory stakeholders meeting was held in Calabar on 20 August 2011 to review the revised document. The UN-REDD Policy Board approved the Nigeria-REDD readiness plan at its Seventh Policy Board Meeting held from 13–14 October 2011 in Berlin, Germany. The Policy Board granted the sum of USD4 million to the Nigeria-REDD Programme. It is clear therefore that local and international NGOs played a critical role in the adoption of REDD+ in Nigeria.

NGO Involvement in Validating the Nigeria-REDD Proposal

Before examining the role of NGOs in launching Nigeria-REDD, it is important to review the rise and importance of conservation NGOs in the CRS. The first forest reserves in Nigeria were created by the British colonial forest policy of 1917, mandating that each colonial province set aside 25% of its forests as reserves (Usman and Adefalu 2010). In Cross

River State, the first forest reserves were created in 1930; logging was forbidden but local people were allowed subsistence use of the forests. These reserves were maintained after independence in 1960, and today there are 14 forest reserves in the state covering 2800 km^2 or 13% of the state's land surface area (Oyebo et al. 2010). The state also hosts the federal government managed Cross River National Park (CRNP), which covers 2,955 km^2 or 18% of the land surface area. Community-managed forests, based on state recognition of customary rights, cover 1632 km^2 or 7% of the land surface area. In all, forests cover what amounts to 40% of CRS's land surface area, which makes up 50% of Nigeria's primary forest (Oyebo et al. 2010).

In the 1980s, the Nigerian Conservation Foundation, with the support of the World Wide Fund for Nature, lobbied the Nigerian government to create the CRNP to protect the critically endangered Cross River gorilla (Harcourt et al. 1989). The WWF initiated a conservation and development project that promised roads, clinics, schools and jobs to win local support for the park. These promises failed to materialise when the European Union (EU) stopped project funding in 1994; this was in protest of the killing of the environmental activist, Ken Saro-Wiwa, by Nigeria's military dictatorship at the time (Ite 1996, 1997, 1998). Local communities were disappointed and turned against the WWF, the park service, the state and the federal government. Several expatriates and local staff who worked for the park project went on to create local NGOs when the project ended (Oyebo et al. 2010). These include Pandrillus, the Centre for Education, Research and Conservation of Primates and Nature (CERCOPAN), the Iroko Foundation, and the African Research Association Managing Development in Nigeria (ARADIN).

In 1999, the British government's Department for International Development (DFID) initiated a community forestry project in CRS (Oyebo et al. 2010). The project led to the creation of community forest groups with membership that included local elites and customary authorities. When the project ended in 2002, there were about 45 community forest groups in CRS, the forestry department in the state Ministry of Agriculture was upgraded to a Forestry Commission with the status of a ministry, and the state had agreed to share timber revenues with local communities where forests had been logged in a 50:50 split (Oyebo et al. 2010). A staff member of an NGO remembers the DFID project with fondness stating that it 'opened our eyes to our rights to the forest and built our capacity to interact with the forestry commission. It was the first time we were treated as equals' (Indigenous NGO staff interview, 2012).

In 2008, the Cross River State governor banned commercial logging and in 2011 appointed an expatriate to head the task force to enforce the ban (Pandrillus 2013). This expatriate runs an NGO involved in biodiversity conservation in CRS. He and his team regularly encounter violent conflict with illegal loggers. This individual carries a firearm always and works with the state's secret service to get the job done. A member of the task force stated that 'since I joined the taskforce my life has not been the same. … I have to be extra careful who I interact with. … [W]hen we seize timber in local communities, the people come out to fight us … when we seize timber coming from Cameroon in the creeks, we also fear for our lives. … [W]e may need armed gun boats' (Timber task force member interview, 2012).

It is commonly held among forest sector actors in the CRS that the commercial logging ban was to please the international community and secure funding for the state's REDD initiative. A member of one of the forest sector NGOs said that 'the log ban cannot be effective … the politicians are involved, those that DFID trained in community forestry are involved. … [T]he governor knows he cannot stop this logging … the ban is to please the international community' (NGO research participant interview, 2012). The commercial logging ban means communities have lost the 50% royalty payments they received from logging revenue collected by the state.

Cross River State forests are a source of livelihood for local people in terms of non-timber forest product harvesting for personal use and commercial sales in local markets (Oyebo et al. 2010). Their elected representatives at the local government level, however, do not have the statutory authority over forestland of more than 50 km^2. Customary authorities exercise some de facto rights over forests, and this is recognised by many community forest groups in the state, but the state governor has the statutory power to ignore customary authority. Cross River State forestry law recognises that local governments can play a role in forest management. So, the Forestry Commission works with local government authorities to set up tree nurseries; the trees are then distributed to farmers and schools for replanting of forests (Oyebo et al. 2010). It is clear then that while NGO and customary authority influence on forest management in the Cross River State has grown since the 1980s, the same cannot be said for local government authorities.

As stated in the previous section, there were two participatory consultative meetings in CRS to validate the Nigeria-REDD proposal. Table 8.2 shows the institutions that attended the first and second Participatory Consultative Meetings.

Table 8.2 Nigeria-REDD participatory consultative meeting participants

Institutions	*First meeting (February 2011)*	*Second meeting (August 2011)*
Federal government	2	0
Cross River State government	37	15
Local government	0	0
Local NGOs based in Cross River State	23	14
National NGOs	2	0
International NGOs	6	1
Participants from local communities	15	36
Media	8	2
Academics	6	4
Banks	2	1
Total	101	73

Source: FME (2012)

What can we learn from Table 8.2? First, NGOs, which we have already noted act as symbolic representatives of local communities through their advocacy work, were prominently in attendance at the Participatory Consultative Meetings. A staff member of the Forestry Commission, which organised the meetings, maintained that NGOs were invited in part because '... the leadership of the commission have come from the NGO sector so we have a very good working relationship with NGOs. ... [T]he people trust the NGOs more than the government' (Forestry Commission staff member interview, 2012). In addition, a staff member of an NGO providing consultancy services to the Forestry Commission stated they were invited because 'we represent local communities, they know us, they trust us more than government ... all we do, we do for local people ... we know the local communities more than anyone else' (NGO staff member of REDD design team interviewed in 2012).

The view that NGOs work for local people has been contested: 'NGOs supporting REDD in Calabar are doing it for what they can get out of it ... they don't have long-term relationship with the people, they go to the villages when money is involved' (Interview of NGO member opposing REDD, 2013). A member of a community organisation added that: 'Na money the NGOs dem want ... if government say no more REDD, make we harvest timber, all of them go say yes [NGOs are after money, if government says no more REDD, let's do logging, they will agree]' (Senior member of local CBO interviewed in 2012). The implication is that the

NGOs that took part in the Participatory Consultative Meetings were doing so for financial gain. It is noteworthy that members of the NGO community in CRS do get recruited as expert advisors and consultants by the Forestry Commission; this is the case for Pandrillus, the service provider or contractor enforcing the CRS government's commercial logging ban.

Second, there were selected individuals from local communities who were invited to the consultative meetings on the basis that they were considered to be descriptively 'typical of the larger class of persons whom they represent' (Mansbridge 1999: 629). These selected individuals are descriptive representatives with no statutory mandate to report back to local communities, and there are no statutory accountability mechanisms through which they could be sanctioned if they chose to stand for personal rather than community interests. Third, elected representatives with a mandate to act for local people were not in attendance at both Participatory Consultative Meetings. The meetings therefore gave room for symbolic and descriptive representation but left out substantive representation of local interests.

What accounts for the absence of elected local government authorities from participatory meetings? Responses from the organisers included: (1) insufficient funds to cover their attendance costs, (2) local officials' are ignorant about environmental issues so do not deserve to be invited, and (3) locals are incompetent administrators. The 18 local government chairpersons, however, have their own budgetary provisions that cover their travel to Calabar for government business; and the local government chairpersons encountered during this study were better educated compared with local chiefs invited to the participatory meetings. Although the charge of administrative incompetence cannot be dismissed (Afrobarometer 2008), the same can be said of state governments and the Nigerian federal government; therefore it is not unique to local governments. As shown earlier in this chapter, local governments in Nigeria are politically weak with respect to decision making about land, and state governments and customary authorities are the key decision makers regarding land use. This explains why they were not considered important enough to be invited to the Participatory Consultative Meetings about the Nigeria-REDD Programme.

Why were NGOs involved in the meetings in Calabar? This in part results from the visibility of forest sector NGOs in the Cross River State. Those in attendance at the Participatory Consultative Meeting included the Nigerian Conservation Foundation, the Rainforest Resource & Development

Centre (RRDC), the African Research Association Managing Development in Nigeria, the NGO Coalition for the Environment, Green Concern for Development (GREENCODE), the Ekuri Initiative, Pandrillus, Concern Universal, One Sky Nigeria and the Wildlife Conservation Society (WCS). These NGOs have a right as members of civil society to participate in the design of Nigeria-REDD just like local individuals and customary authorities. Still, NGO involvement in validating the Nigeria-REDD proposal became a replacement action for local government authorities. The next section looks at the limitations of NGOs as political representatives of local people.

Limitations of NGOs Involved in Launching Nigeria-REDD

Local and international NGOs played a critical role in the adoption of REDD+ in Nigeria. They advocated for the adoption of carbon forestry by the CRS government as a biodiversity-friendly means of generating revenue from the state's forest resources. These NGOs also were involved in the design of the Nigeria-REDD readiness proposal, and in the Participatory Consultative Meetings to validate the REDD+ proposal document. Even though these NGOs do make claims to represent local communities and do speak for local people through their advocacy activities, their representative claim is questioned by some at the local level in CRS who maintain that NGOs are acting for financial gain rather than for the public good.

Evidence of this is the fact that these NGOs are willing to serve as paid consultants to the CRS government, working to implement state government policy. In accepting government funds, can these NGOs still serve as watchdogs pressuring government to formulate and implement policies in a manner that is responsive to the needs of local people? In the Cross River State, NGOs that have expressed the view that the state should not adopt REDD+ find that they are labeled 'enemy of progress' by the Forestry Commission (Oluwatoyin Abiola, pers. comm. 31 August 2013),[2] as well as kept away from the REDD+ process.

The Forestry Commission ensures that only those NGOs that agree with its preferences get invited to its REDD+ consultative meetings. Furthermore, in a candid interview with a senior staff member of an NGO invited to the participatory meetings, he stated that 'we were informed of the [consultative] meeting by text message by 11pm the night before, we arrive[d] in the morning … not well prepared to comprehensively engage with the UN-REDD people' (NGO leader interview in Calabar, 2012). A member of another NGO said: '... [W]e missed the participatory consultative meetings, we were informed by text at midnight ... no way for us to pass the message to our local communities, we feel frustrated' (Frustrated NGO staff member interview in Calabar, 2012). These individuals were of the view that they were invited to the meetings by the Forestry Commission so that their presence would legitimise the proceedings, not so that they could critically engage and debate the contents of the REDD+ readiness proposal document. Local NGO members realised that they were being used as pawns, but could not challenge the Forestry Commission if they wanted to continue as consultants and to receive support for their work from the commission.

Although local NGOs find that they may lose an important source of funding and bureaucratic support if they critically oppose the state government, local government authorities through the Association of Local Government of Nigeria (ALGON) have been challenging their subjugation by state governors in the media, in the courts, in Nigeria's House of Representatives and Senate (CLGF 2011; Akpan and Ekanem 2013; Vanguard 2014; Olaniyi 2015; *The Guardian* 2015). Local governments can challenge higher level political authorities in a way that local NGOs are not able to because they have a constitutional mandate and a political mandate from their electorate to be responsive to the social development needs of their local communities. It is on the strength of their constitutional and political mandates that local governments challenge the arbitrary overbearing decisions of state governments. Local governments therefore can make demands for the enfranchisement of local people whom they are elected to represent in a way that NGOs cannot hope to do. This limits the effectiveness of local NGOs as representatives and development actors at the local level.

Despite the administrative shortcomings of local government authorities, talk of replacing local governments with NGOs (UNDDSMS and UNDP 1996) happens to be wishful thinking. The NGOs at their best are advocates for the locals in subnational to international decision-making arenas not accessible to locals, and where government representatives may not adequately speak up for the interests of local people. Most NGOs enjoy a high degree of trust from local communities because they (local communities) are not as aware of governance challenges among NGOs (Holloway 1999; Chapin 2004) as they are about the governance weaknesses of local government authorities. Still, devolving power from national governments to NGOs to govern local territorial spaces will concentrate power in the hands of experts with no accountability relationship with the people—a recipe for authoritarianism.

In the case of the NGOs involved in validating the Nigeria-REDD Programme, they have not been devolved power to govern local administrative spaces in Nigeria, but they have been complicit in the symbolic political actions of the CRS Forestry Commission. The way they received invitations to the Participatory Consultative Meeting at midnight for a morning meeting, allowing no time to prepare for deliberations, and the knowledge that the Forestry Commission is intolerant of dissenting voices, raised suspicions that they were invited so that their presence would contribute to legitimising the Nigeria-REDD readiness process in the public's eyes, and in fulfilment of the requirements set by the UN-REDD. It was felt that the Forestry Commission was just creating a show of compliance—that is, a symbolic action instead of a substantive response to UN-REDD requirements.

Based on Bluhdorn's symbolic politics schema (2007), discussed earlier in this chapter, the Forestry Commission opted for a replacement action instead of a substantive response to the UN-REDD. Bluhdorn study (2007) provided two major reasons and four plausible explanations as to why an institution would choose symbolic replacement action over substantive response. Table 8.3 is from his article but with the irrelevant parts left out.

The Forestry Commission was not under contradictory citizen pressure so option B1.1 is not in consideration; so too option B2.1 because the

Table 8.3 Symbolic politics as replacement actions

Symbolic politics			
Replacement action (B)			
Best possible option (B1)		*Deliberate strategy (B2)*	
In light of fragmented and contradictory citizen interests	In light of imperatives emerging from the political system	To temporarily delay the uncomfortable consequences of effective action	To deceive the public and pursue a secret elite agenda
(B1.1)	(B1.2)	(B2.1)	(B2.2)

Source: Adapted from Bluhdorn (2007)

Forestry Commission was not against effective action for fear of negative consequences. That leaves B1.2 and B2.2. Was there a secret agenda in play with respect to the involvement of the CRS government and its Forestry Commission in Nigeria-REDD? This is not plausible because the Forestry Commission made it known to all that they were interested in REDD+ for the income it would generate for the state.

This leaves option B1.2; that is, the Forestry Commission opted for symbolic action as the best possible option considering the pressure emerging from the political system. The Forestry Commission was under pressure from the state governor to act quickly with respect to validating the Nigeria-REDD proposal document. It was afraid that local NGOs, being the most informed institutions invited to the participatory meetings, could delay the process if they were able to fully engage in deliberations about the proposal document. The commission therefore acted proactively to disable the ability of local NGOs to debate the merits of the Nigeria-REDD proposal document (Oluwatoyin Abiola pers. comm., 2013).

Concluding Remarks

The NGOs are key actors in the design and implementation of environmental conservation initiatives around the world. They partner with donors, international organisations and governments to implement conservation and/or conservation with development projects and programmes. Sometimes NGOs also work as contracted service providers for larger NGOs, donors and governmental institutions. Several NGOs have become so involved in the design and implementation of environment and development initiatives that members have opined that they could even replace local government authorities—the substantive political representatives of local people (UNDDSMS and UNDP 1996).

This chapter has examined the strengths and limitations of conservation NGOs as the political representatives of local people. It asks: Where or from whom do NGOs derive their authority and legitimacy to speak for and execute projects addressing local development needs? It analyses the role of conservation NGOs in local environment and development initiatives informed by the theory of political representation. It presents results from a study of the role of conservation NGOs during the consultative process in CRS that accompanied the design of the Nigeria-REDD readiness proposal, under the supervisory oversight of the UN-REDD Programme. Cross River State is the pilot site for forestry activities included in the Nigeria-REDD readiness proposal.

This chapter shows that NGOs dominate the forest conservation sector in CRS, with some working as allies of the state government. These selected allied NGOs were invited by the Cross River State Forestry Commission to participate in the consultative process to validate the design of the Nigeria-REDD readiness proposal. The NGOs were required by the UN-REDD to be part of the consultative process because they self-proclaimed to be and are recognised as advocatory or symbolic representatives speaking for local communities. Customary authorities with their de facto rights over land also were invited to participate in the consultative process. Yet, local government authorities, the substantive political representatives of local people, were excluded from the process because they were deemed politically weak—their political authority having been curtailed by the state government.

The CRS Forestry Commission invited local NGOs to the meetings as a show of having implemented a comprehensive participatory process, when in reality these NGOs barely had time to review the Nigeria-REDD readiness proposal before they were asked to give it their approval. Political pressures from the state government on the Forestry Commission to hasten the participatory validation process influenced its decision to opt for symbolic validation rather than a substantive process involving detailed review of the Nigeria-REDD proposal by informed local NGOs in the Cross River State.

Even though the NGOs who were part of the consultative participatory process realised they were pawns, they felt powerless to question the Forestry Commission. These NGOs needed the bureaucratic support of it to facilitate their activities, as well as to benefit from consultancy opportunities at the Forestry Commission. In contrast, local government authorities, while subjugated by state governors, fought back through their umbrella association, ALGON, using the media, the courts, and the legislature to fight against the overbearing actions of state governors. Their constitutional and political mandate, plus their limited but statutory fiscal autonomy, empowers them to challenge the higher level government structures. On the other hand, NGOs in Nigeria lack this type of constitutional, political and fiscal foundation, so this makes them hesitant to confront governmental authorities.

Notes

1. This is based on an exchange rate of 160.50 Naira to USD1 as of the time of the field research.
2. This is not the real name of the research participant because the individual's identity must be kept confidential.

References

Adesopo, A. 2011. Re-examining the Failing Inter-governmental Fiscal Relations and Sustenance of Nigerian Federation: An Empirical Study. *Asian Social Science* 7 (10): 107–121.

Afrobarometer. 2008. *Public Opinion and Local Government in Nigeria, 2008.* Afrobarometer Briefing Paper No. 53.

Akpan, F., and O. Ekanem. 2013. The Politics of Local Government Autonomy in Nigeria Reloaded. *European Scientific Journal* 9 (35): 193–205.

AKSG (Akwa Ibom State Government). 2012. 76 Oil Wells—How Cross River State Went to Court and Lost. http://www.africanspotlight.com/2012/07/23/76-oil-wells-how-cross-river-state-went-to-court-and-lost/. Accessed 28 June 2013.

ARD. 2001. *Nigeria: Local Government Assessment*, Report prepared for USAID. Burlington: ARD, Inc.

Bluhdorn, I. 2007. Sustaining the Unsustainable: Symbolic Politics and the Politics of Simulation. *Environmental Politics* 16 (2): 251–275.

Brockington, D., and K. Scholfield. 2010. The Work of Conservation Organisations in Sub-Saharan Africa. *Journal of Modern African Studies* 48 (1): 1–33.

Brysk, A. 1995. 'Hearts and Minds': Bringing Symbolic Politics Back In. *Polity* 27 (4): 559–585.

Chapin, M. 2004. A Challenge to Conservationists. *World Watch Magazine* 17 (6): 17–31.

CLGF (Commonwealth Local Government Forum). 2011. News Release—CLGF Pledges to Help Strengthen Local Government in Nigeria. http://www.clgf.org.uk/whats-new/news/clgf-pledges-to-help-strengthen-local-government-in-nigeria/. Accessed 28 June 2013.

CRS (Cross River State). 2008. Communique of the Stakeholders' Summit on the Environment Organized by the Government of Cross River State from the 25–28 June 2008. Calabar, Nigeria. http://tropicalforestgroup.org/wp-content/uploads/2012/10/crsoutcomesdOC.pdf. Accessed 22 Feb 2017.

Diejomaoh, I., and E. Eboh. 2010. Local Governments in Nigeria: Relevance and Effectiveness in Poverty Reduction and Economic Development. *Journal of Economics and Sustainable Development* 1 (1): 12–28.

Dryzek, J.S., and S. Niemeyer. 2008. Discursive Representation. *American Political Science Review* 102 (4): 481–493.

Edelman, M. 1985. *The Symbolic Uses of Politics*. Champaign: University of Illinois Press.
FGN (Federal Government of Nigeria). 1978. *Nigeria Land Use Act 1978*. Abuja: Federal Government of Nigeria.
———. 1999. *Constitution of the Federal Republic of Nigeria*. Abuja: Federal Government of Nigeria.
Filou, E. 2010. Odigha Odigha: Speaking Truth to Power. http://www.ecosystemmarketplace.com/pages/dynamic/article.page.php?page_id=7497§ion=news_articles&eod=1. Accessed 28 Apr 2016.
———. 2011. Nigerian State Sets REDD Pace for Entire Continent. http://www.ecosystemmarketplace.com/articles/nigerian-state-sets-redd-pace-for-entire-continent/. Accessed 28 Apr 2016.
FME (Federal Ministry of Environment). 2012. *Nigeria's REDD+ Readiness Programme (2012–2015)*. Abuja: Federal Ministry of Environment.
Forest Trends, the Katoomba Group and UNEP (United Nations Environment Programme). 2008. *Payments for Ecosystem Services: Getting Started—A Primer*. Washington, DC: Forest Trends and the Katoomba Group.
Fraser, N. 2005. Reframing Justice in a Globalizing World. *New Left Review* 36: 69–88.
Global Witness. 2011. Review of Nigeria's NPD Submitted to the 7th Policy Board Meeting of the UN-REDD Programme. https://www.globalwitness.org/sites/default/files/nigeria%20npd.pdf. Accessed 22 Feb 2017.
Grunebaum, J.O. 1981. What Ought the Representative Represent? In *Ethical Issues in Government*, ed. N.E. Bowie, 54–67. Philadelphia: Temple University Press.
Harcourt, A.H., K.J. Stewart, and I.M. Inahoro. 1989. Nigeria's Gorillas. *Primate Conservation* 10: 73–79.
Holloway, R. 1999. NGOs: Loosing the Moral High Ground-Corruption and Misrepresentation. In: the 8th International Anti-Corruption Conference (IACC). http://8iacc.org/papers/holloway.html. Accessed 22 Feb 2017.
Ite, U.E. 1996. Community Perceptions of the Cross River National Park, Nigeria. *Environmental Conservation* 23 (4): 351–357.
———. 1997. Small Farmers and Forest Loss in Cross River National Park, Nigeria. *Geographical Journal* 163 (1): 47–56.
———. 1998. New Wine in an Old Skin: The Reality of Tropical Moist Forest Conservation in Nigeria. *Land Use Policy* 15 (2): 135–147.
Kateb, G. 1981. The Moral Distinctiveness of Representative Democracy. *Ethics* 91 (3): 357–374.
Keck, M.E., and K. Sikkink. 1999. Transnational Advocacy Networks in International and Regional Politics. *International Social Science Journal* 51 (159): 89–101.
Konings, P. 2011. Settling Border Conflicts in Africa Peacefully: Lessons Learned from the Bakassi Dispute Between Cameroon and Nigeria. In *Land, Law and Politics in Africa: Mediating Conflict and Reshaping the State*, ed. J. Abbink and M. de Bruijn, 191–210. Leiden: Brill.

Lombardo, E., and P. Meier. 2014. *The Symbolic Representation of Gender: A Discursive Approach*. Farnham: Ashgate Publishing Company.

Manin, B., A. Przeworski, and S. Stokes. 1999. Introduction. In *Democracy, Accountability and Representation*, ed. A. Przeworski, S. Stokes, and B. Manin, 1–26. Cambridge: Cambridge University Press.

Mansbridge, J. 1999. Should Blacks Represent Blacks and Women Represent Women? A Contingent 'Yes'. *The Journal of Politics* 61 (3): 628–657.

Mill, J. S. 1861. *Considerations on Representative Government*. London: Parker, Son, and Bourn. https://archive.org/stream/considerationso04millgoog#page/n16/mode/2up. Accessed 22 Feb 2017.

Miller, H.T. 2012. *Governing Narratives: Symbolic Politics and Policy Change*. Tuscaloosa: University Alabama Press.

Montanaro, L. 2012. The Democratic Legitimacy of Self-Appointed Representatives. *The Journal of Politics* 74 (4): 1094–1107.

Nuesiri, E.O. 2003. Conservation and Resource Use in Cross River—Takamanda Forest Complex of Cameroon and Nigeria. Report Submitted to Fauna and Flora International (FFI). https://www.academia.edu/4356469/Report_on_conservation_and_resource_use_in_Cross_River_-_Takamanda_forest_complex_of_Cameroon_and_Nigeria. Accessed 22 Feb 2017.

———. 2015. *Representation in REDD: NGOs and Chiefs Privileged over Elected Local Government in Cross River State, Nigeria*, RFGI Working Paper No. 11. Dakar: CODESRIA. https://portals.iucn.org/library/node/45265. Accessed 22 February 2017.

Oates, J.F., R.A. Bergl, and J.M. Linder. 2004. *Africa's Gulf of Guinea Forests—Biodiversity Patterns and Conservation Priorities*, Advances in Applied Biodiversity Science Number 6. New York: Wildlife Conservation Society.

Oates, J.F., J. Sunderland-Groves, R. Bergl, A. Dunn, A. Nicholas, E. Takang, F. Omeniet, et al. 2007. *Regional Action Plan for the Conservation of the Cross River Gorilla (Gorilla Gorilla Diehli)*. Arlington: IUCN/SSC Primate Specialist Group and Conservation International. https://portals.iucn.org/library/sites/library/files/documents/2007-012.pdf. Accessed 22 Feb 2017.

Olaniyi, B. 2015. Rivers Local Governments of Controversy. http://thenationonlineng.net/rivers-local-governments-of-controversy/. Accessed 18 Jan 2016.

Olubusoye, O.E., and T.D.T. Oyedotun. 2012. Quantitative Evaluation of Revenue Allocation to States and Local Governments in Nigeria (1999–2008). *European Scientific Journal* 8 (3): 224–243.

Omar, M. 2012. Ensuring Free, Fair and Credible Elections in Local Governments in Nigeria. *Developing Country Studies* 2 (11): 75–81.

Oyebo, M., F. Bisong, and T. Morakinyo. 2010. *A Preliminary Assessment of the Context for REDD in Nigeria*. Report Commissioned by Nigeria's Federal Ministry of Environment, the Cross River State's Forestry Commission and United Nations Development Programme. http://www.unredd.net/index.php?option=com_docman&task=doc_download&gid=4129&Itemid=53. Accessed 22 Feb 2017.

Pandrillus. 2013. Peter Jenkins Installed as Chairman of the Cross River State Anti-deforestation Task Force (ATF). http://www.pandrillus.org/news/peter-jenkins-installed-as-chairman-of-cross-river-state-anti-deforestation-taskforce/. Accessed 28 June 2013.

Pitkin, H.F. 1967. *The Concept of Representation*. Berkeley: University of California Press.

Rehfeld, A. 2006. Towards a General Theory of Political Representation. *The Journal of Politics* 68 (1): 1–21.

———. 2011. The Concepts of Representation. *American Political Science Review* 105 (3): 631–641.

Ribot, J.C., E. Barrow, and E. Sall. 2010. Responsive Forest Governance Initiative (RFGI): Project Summary and Introduction. http://www.codesria.org/IMG/pdf/RFGI_Summary.pdf. Accessed 22 Feb 2017.

Runciman, D. 2007. The Paradox of Political Representation. *Journal of Political Philosophy* 15: 93–114.

Saward, M. 2006. The Representative Claim. *Contemporary Political Theory* 5 (3): 297–318.

———. 2008. Representation and Democracy: Revisions and Possibilities. *Sociology Compass* 2: 1000–1013.

Silveira, S. 2001. The American Environmental Movement: Surviving Through Diversity. *Boston College Environmental Affairs Law Review* 28 (2): 497–532.

TFG (Tropical Forest Group). 2011. *Reducing Emissions from Deforestation*. Nigeria: Cross River State. http://www.tropicalforestgroup.org/logging-moratorium-cross-river-state-nigeria/. Accessed 28 Apr 2016.

The Guardian. 2015. Wike, ALGON Disagree over Right to Borrow from Banks. http://guardian.ng/news/wike-algon-disagree-over-right-to-borrow-from-banks/. Accessed 28 Apr 2016.

UNDDSMS (United Nations Systems Department for Development Support and Management Services) and UNDP (United Nations Development Programme). 1996. *Local Governance: Report of the United Nations Global Forum on Innovative Policies and Practices in Local Governance*. Gothenburg, Sweden, 23–27 September. https://publicadministration.un.org/publications/content/e-library.html. Accessed 22 Feb 2017.

Urbinati, N., and M. Warren. 2008. The Concept of Representation in Contemporary Democratic Theory. *Annual Review of Political Science* 11: 387–412.

USAID. 2010. *Property Rights and Resource Governance—Nigeria, USAID Country Profile*. Washington, DC: USAID.

Usman, B.A., and L.L. Adefalu. 2010. Nigerian Forestry, Wildlife and Protected Areas: Status Report. *Tropical Conservancy* 11 (3–4): 54–62.

Vanguard. 2014. LG Autonomy: ALGON Lauds Stakeholders. http://www.vanguardngr.com/2014/11/lg-autonomy-algon-lauds-stakeholders/. Accessed 28 Apr 2016.

Wintrobe, R. 1998. *The Political Economy of Dictatorship*. Cambridge: Cambridge University Press.

CHAPTER 9

Misreading the Conservation Landscape

Kent H. Redford

Introduction

In a seminal book, social scientists Fairhead and Leach (1996) demonstrated how forestry officials and conservationists misread the effects of local peoples on the extent of forest cover in the Republic of Guinea. Using painstaking fieldwork incorporating the biological, social and historical sciences, they concluded that instead of being forest destroyers, these people were responsible for increases in forest cover—their role had been completely misjudged. Fairhead and Leach's approach is exemplary in using careful scholarship and informed interpretation to understand local practises, beliefs and behaviours. This is an approach to inquiry that appears to have been forgotten by some of the social science community writing about the practise of conservation.

I write here as a practising conservationist trained as a biologist. Over the last few decades I have learned a great deal about my own discipline and practise through reading the works of, and collaborating with, social scientists. Building on my experience, in this chapter I make the case for a more strategic, systematic openhanded collaboration between conservation

K.H. Redford (✉)
Archipelago Consulting, Portland, ME, USA

P.B. Larsen, D. Brockington (eds.), *The Anthropology of Conservation NGOs*, Palgrave Studies in Anthropology of Sustainability, DOI 10.1007/978-3-319-60579-1_9

practitioners and social scientists, with a focus on anthropology, political science and geography. The objective of such collaboration would be to create a resilient practise that conserves the world's biodiversity while respecting and empowering people.

Such a practise will benefit not only from the direct application of improved practises but also, indirectly, from the careful study of practise, practitioners and the culture of both. I suggest two things: (1) that some of the methods, the use of which has led to so many significant advances in social science, are not being used to best advantage in studying conservation; and (2) that careful empirical work on conservation practise by social scientists can better inform the practise of conservation.

We are on a common journey shaped by the growing power of human influence on both nature and the people whose lives are most directly intertwined with the natural world. Long-standing differences that typify those of us who came to this field decades ago are not necessarily shared by a younger generation, trained to think across disciplinary boundaries and with evolving sets of values. Here I hope to help lay the groundwork for a more effective collaboration between disciplines, between discipline and practise, and in the training of the next generation of those interested in the fate of the natural world (Mascia et al. 2003). As a way of advancing this agenda, I provide two profiles.

First is a summary of the ways in which I think social science work has already improved conservation practise. Second is a set of generalisations made by some social scientists about the practise of conservation that are incorrect or incomplete. Sometimes these are made explicit and sometimes they remain implicit. In both cases they deserve considerably greater examination and, if reconsidered, may further strengthen the contribution that social science could make to conservation practise. The first profile shows the power of the partnership I advocate and the second points to what I think are the most important things to address to better inform collaborative efforts between social scientists and conservation practitioners. What is here is not meant to be exhaustive but illustrative. I am sure that others could make itemisations different from mine.

Not all social sciences are the same and neither are all social scientists. The same is true for conservationists and this author is aware that what I call 'conservation' is a diverse family. Equally true is the gradual dissolving of the distinction between us—the conservationists—and them—the social scientists. Yet despite this, there remains an apparent anticonservation orthodoxy in some of the social science literature that, in my opinion, is inhibiting improved conservation practise, greater collaboration, greater incorporation of key social science values and methods into conservation practise, and improved interdisciplinary training of young social scientists and conservation practitioners. I hope both to applaud what I think are some of the contributions the social sciences have made to conservation and to highlight some of the generalisations that I think are impeding further constructive engagement. Extensive citations have not been provided because I wish to focus on a general reading of the literature.

Ways in Which the Social Sciences Have Improved Conservation Practise

Showing That Conservation Is More Than Biology

The contemporary practise of conservation was born out of conservation biology, where its roots have remained. Most practising and academic conservationists were trained in various disciplines of the natural sciences. Conservation organisations have promoted science-driven approaches that have largely ignored the social sciences. Perhaps it was an act of hubris on our part to assume that such training equipped us to practise conservation, but we were the only ones stepping forth, and so we did. As a result, conservation practise has been based on biocentric values and assumptions, privileging natural science views of both problems and solutions.

But then, we have learned that conservation practise is politics—a public discussion about the allocation of resources—and it has taken repeated assertion of this fact for the discipline to admit that social science and social scientists are key to implementing and improving the practise of conservation. Much has been learned from social science examinations of the nature of power, cross-scale partnerships, institutional structure and environmental governance. We need to better understand the social sciences, we need to appreciate what these disciplines offer, and we need social scientists to be more involved in conservation.

Remembering History I

Most conservationists were taught that history is 'evolution'. We suffer from a remarkable inability to think about ecological history, let alone human history. The first history we learned has been termed 'historical ecology', which has been trying to teach us that humans have been a driving ecological—and even evolutionary—forces for a long time, and that notions of pristine nature are not only inaccurate but also have stood in our way as we try to achieve conservation in a human-dominated world.

Remembering History II

The second type of history we have had to learn is our own history. We have ignored, forgotten or falsely constructed the historical legacy of conservation and then been puzzled that so many of our actions have been rejected by those who not only remember the history but also have been victims of it. The imperialist roots of conservation in many parts of the developing world, and the forcefully imposed nature of conservation there, have been woven into the fabric of contemporary feelings about conservation practise. This fact was unknown or ignored until impressed on us by social scientists. Recognising these historical sensitivities has laid the groundwork for improved practise.

Acknowledging People: Conservation Is Practised by People with a Mixture of Ethical Positions

It is fair to characterise much of conservation, at least in the developed world, as being firmly rooted in a biocentric position. We often see humans as threats to the biological systems we champion. Whereas it is true that the current dismal state of the biosphere is due in large part to the accumulated human impact, it is equally true that any success in altering this will require human action. We have been chastised by social scientists for talking about humans only as threats—a persuasive admonition that has contributed to a gradual move toward viewing humans as legitimate elements in nature and as an explicit part of the solutions to conservation problems. Social scientists have been largely responsible for bringing the importance of understanding, and more fairly balancing, the human costs and benefits of conservation to our attention.

The Culture of Conservation and the Culture of Conservationists

Many conservation practitioners were content to work in the ways we were taught in school: eyes on the immediate path in front of us. We were not reflective practitioners, being too busy addressing the extinction crisis and rushing from one grant, one publication, one project and one challenge to the next to take the time to reflect on our work and our successes and failures. It took criticism from many, including social scientists, for us to recognise that we were both part of the problem and part of the solution; and, most important, part of a culture that affected the ways we thought and acted and the outcome of our practise. Recognising the cultural contexts—institutionally, nationally and within our discipline—led to a clearer recognition of our assumptions and therefore of our actions. This was a vital first step in helping to understand what we are doing and in understanding how others view our work. From such an understanding, we hope to bring greater support for conservation.

Valuable Social Research and Interdisciplinary Collaboration

The social sciences have developed methods to study diverse aspects of human societies. This has led to rich findings with relevance to conservation and has included topics such as resource tenure, governance, institutions, power, adaptive capacity, human rights and gender. Those interested in more effective conservation can learn from social science research and/or employ social science tools to improve their understanding and practise. We also have learned that inadequate or cursory social research by untrained biologists may be worse than no research at all, and that careful, respectful interdisciplinary work has enormous potential and can even develop new fields of enquiry (McClanahan et al. 2009).

Conservation Through a Glass, Darkly: Remaining Challenges for Collaboration

The second profile contains a set of generalisations made by some social scientists about the practise of conservation that are incorrect or incomplete. These are not always made explicit but often are. In either case, I contend that they warrant more careful examination and that this could lead to both a better understanding of the practise of conservation and a better-informed collaboration between conservation practitioners and the social sciences.

Conservation Is One Thing, with One Practise and One Set of Practitioners

Conservation is written about as if it exists as a monolithic practise with identical practitioners. This claim allows a Manichean depiction of the world, with the negative role almost inevitably assigned to conservation:

portrayed as undermining the livelihoods of the poorest in pursuit of strictly biocentric aims. There are, however, significant differences in conservation organisations, practises and individuals that vary by personal and institutional values, scale, strategies, nationalities, politics, experiences and priorities. Arguments among the conservation community in settings, such as the World Parks Congress or the Convention on Biological Diversity meetings, are clear proof of the lack of a single position.

The result is a tremendous richness within the practise of conservation that prevents simple generalisation. For example, the Wildlife Conservation Society (WCS), the organisation for which I used to work, operates in 61 countries with almost 2000 staff members and more than 300 projects with a wide range of traditions and values. Yet, oversimplifications continue to obscure the diversity of conservation, making difficult the collaboration necessary for more effective conservation and potentially impeding the incorporation of more social science practises into our work. Generalities obscure the strength of debate within conservation—for example, between concerns about living diversity in all forms, versus rare or charismatic species, versus the capacity of ecosystems to supply services to humankind. This variety of concerns leads to various priorities, pursued often by different organisations. Such differences matter, and social scientists could help analyse and explain them.

Community-based conservation, for example, is thought by many conservationists to be largely about development and sustainable livelihood promotion and thus not part of the mainstream of conservation. Yet, the definition of what is 'in' and what is 'out' is itself a matter of contention, voice and power. Collaboration built on an understanding of such variations within conservation, its sources, and its consequences would provide a grounded understanding of what would help us in our practise and more accurately portray the work of conservation practitioners.

Conservation Occurs in Epochs

The reading of the history of contemporary conservation by some social scientists is frequently shallow. Usually tossed off in a few sentences or a few paragraphs, many authors tell a neat story about the history of conservation that has one strategy replacing the one preceding it with artificial regularity. In some interpretations, ecoregional-based conservation is said to have been developed in response to the stated failures of community conservation. In others, community-oriented projects have been displaced by payment-based schemes.

Although these statements sound compelling, they are neither documented nor appear to have been informed by consultation with those of us who have been involved in such changes. For example, as an early advocate for ecoregional conservation, I can say that the efforts with which I was involved had absolutely nothing to do with community conservation. Yet no one has asked. We would benefit from serious attention to detail; informant-based information; and variation between places, organisations, and individuals—the sort of thing that social scientists have been trained to do. Those who examine this history need to recognise that various approaches rise and fall but never disappear.

All serious conservation portfolios contain projects that mix and match diverse strategies. The modern practise of conservation does change, but its history has not yet been accurately written and the bricolage that currently masquerades as the telling of conservation history is a further deterrent to mutual understanding and collaboration; however, it does purport to define a history that those of us involved know is not correct.

All Parks Are the Same, and They Are Only About the Protection of Biodiversity

The park often is featured as a stereotyped entity. This archetypal park is in the tropics; people were ejected at the time of its creation and local people are prohibited from using it. It is visited by wealthy tourists from the developed world, run by an autocratic government, enabled by a wealthy international conservation organisation, and devoted to the maintenance of large animals in unnatural densities through practises copied from the original inhabitants. That such places exist is indisputable. But all parks are hardly like this. This unfortunate tendency to lump all categories of protected areas (PAs) under the single appellation of 'parks' and then to define all parks as only about protection of nature is incorrect.

As of 2005, there were more than 114,000 legally gazetted PAs registered in the International Union for Conservation of Nature (IUCN) database (Chape et al. 2008). These fall into six major categories that vary from strict nature reserves to those that protect sustainable use of natural ecosystems. Only 50% of the total land area under any legal protection is designated for strict protection (Categories I–IV; Jenkins and Joppa 2009), with 41.4% in areas designed for sustainable use—most with legal human habitation within the PA (Categories V–VI; Chape et al. 2008). In Italy, for example, the majority of the land within protected areas remains under private ownership, with sanctioned agriculture and grazing (Parks.it 2017).

Additionally, recent changes in the IUCN's classification of PAs have created a matrix with multiple governance types ranging from national government to community conserved areas (Dudley 2008; Dudley et al. 2010). Clearly, there is no longer such a single entity as a 'park' because there is such a single entity as a community, as social scientists have demonstrated (Agrawal and Gibson 1999). If social scientists differentiate more obviously between various kinds of protected areas, it would sharpen their analysis and critique and improve their impact among conservation professionals.

Conservation Is Only About Parks

Conservation often is equated only with parks, and the assumption is then made that support for parks means that practitioners think that conservation outside them is unimportant. Although most conservation practitioners think that PAs are a vital tool in achieving biodiversity conservation, there is a widespread belief that this is by no means the only tool that should be used. Protected areas are not islands and never were; even focussing on them alone requires working beyond their boundaries. Approaches to conservation that predate modern approaches to protected areas are rich and deep, and there is extensive literature that addresses these approaches.

Conservation has never been only about parks, and there has been more work done on conservation outside parks than inside them. Such conservation tools include land-use planning, resource harvesting, sustainable livelihood promotion, poverty alleviation, and human consumption, and there is abundant literature surrounding each of them in both natural and social science journals. Social science critiques of conservation often underrepresent this work.

Conservation Is Only a Project of the Developed World's Elite Inhabitants

It is true that the contemporary conservation movement has been organised and promoted primarily by elite inhabitants of the developed world and supported by institutions staffed, funded and/or owned by members of the elite. Nonetheless, the claim that these are the only partisans of conservation is to elide the genuine, widespread and long-standing interest in conservation among many groups of people all over the world.

By defining the contemporary conservation movement as only park-based and locating the origin myth in Yellowstone National Park, the argument has become orthodoxy and thus created conditions that deny agency to many groups. Included among them are the First Nations in Canada and their emerging tribal parks movement; the ancient origin of Mongolian PAs, the increasing recognition of indigenous and community-conserved areas, and the sacred connection to protected areas by many of the world's religions. It also denies to citizens of developing nations the right to claim they too are conservationists—that is, not as lackeys of the West or because of colonial determinism. Instead, it is because they have become convinced of the importance of conservation in this changing world or wish to defend long-standing cultural practises promoting conservation.

Women's groups, schools, religious communities, villages, provinces, ethnic groups and nations have demonstrated their commitment to conservation. Conservation is not now, if it ever was, solely an imposed, top-down belief system of elite Western citizens, although this version has had an undeniable negative impact in its execution. Social science could contribute greatly to the understanding of the diverse ways people engage with the ideas and practises of conservation.

Conservation in General and Parks in Particular Are Bad for Local People

Although protected areas have been shown in some cases to be sources of major problems for local residents, this is not always the case, and practises are changing fast. There are a significant and growing number of cases ranging from the Chaco of Bolivia to western Canada where the establishment of parks is promoted by local people. Park establishment has been used by them to provide tools to help control large-scale mining and agricultural conversion, to help conserve ways of life, and to bring international attention to their culture and problems.

Large PAs declared by local people now exist on every continent, and many of them are managed for the presence of local cultural and economic activities. Lack of acknowledgment or study of these situations by social scientists writing about conservation denies the complexity that is an inevitable part of conservation as well as the complex ways, positive and negative, that people use PAs and, in turn, are affected by them.

Marketing of Nature Is a Modern Phenomenon

Slumbering under the arguments of many who write about trends in conservation is an assumption that in the past nature and natural resources, and the people who lived with them, were not subject to local, regional and international markets. This is an ahistorical rendering of the facts because there has been international trade in products (e.g., obsidian, frankincense, musk, animals for the Roman circuses and ivory) for thousands of years, with this trade affecting the resources involved. Trade influenced patterns of human settlement, resettlement, enslavement and occupation long before modern capitalism. Nature also was commodified as hunting, scenery and gardens for thousands of years across most continents.

The interplay between nature, commerce and humans is a long, complicated one that denies simple generalisation and informs current interests in creating novel markets for nature. There is a need for research into the historical understanding of the human exploitation of wild species under various economic conditions and over time. The emerging body of work on novel markets and links to globalisation is important but needs to be tied to earlier patterns of buying and selling nature.

Conservation in One Part of the World Is the Same as in Other Parts of the World

Much of the most influential literature in this area has been authored by social scientists with experience in eastern Africa but with important contributions from Southeast Asia and a few other locations. The experiences from this limited range of places has been generalised to conservation as a whole. Conservation experience from the United States (US) and Europe appears much less frequently except when it supports the critique of US national park establishment related to Native Americans.

Particularly notable for its low profile in the discussions referenced here is the experience from Europe, where the conservation narrative would seem to differ significantly from much of the rest of the world, and Latin America, where experiences with Native American communities have produced a variety of outcomes. This is not to say that the case made for the East Africa conservation story is not true but that there is no single truth even from this region, and extrapolation can be done only with meticulous care. Social science could explore how conservation differs in various geographical and social settings with different histories and ecologies.

What Is Published by Conservationists Is Representative

There is a tendency to assign to the entire conservation community the opinions of a few passionate individuals. Other perspectives by these same authors and by many others are passed over in the pursuit of a simple characterisation of the whole community. The result has been a synecdoche—a misreading that appeals to other like-minded authors—that becomes self-perpetuating, and through entrainment of the argument, increasingly inaccurate. There is a great diversity of opinion in the community of conservation practitioners and no one speaks for the whole community.

Added to this is the fact that most conservation practitioners do not publish in peer-reviewed literature, if at all. Adjudging what they know and what they think by referring only to the published literature is fraught with dangers. Instead, careful work with this community, using the time-tested tools of social science, will provide a treasure trove of richness, contradiction, myth and truth—the perfect material for the social scientist.

Discussion

It will be easy for some to read this chapter and dismiss it as yet another attempt by a beleaguered inhabitant of a fortress to strengthen his or her position. But then, this would be a misreading of my intentions. I was trained to think that all I needed to know to save the world was biology, and it took decades for me to understand how wrong I was. This author has come to understand that conservation is (1) about politics and power, (2) about people and societies and history, (3) about morals and values, and (4) about how people view the world and make decisions—all are fields of study in the social sciences.

Self-styled as a natural science, conservation is rooted in a set of values and assumptions that privilege the natural science perspective and uses a language to describe itself, its aims, and the rest of the world that reinforces this centrality. Nevertheless, conservation is not only a natural science, but it also is a blend of natural and social science, with a large dose of art. People trained in either of these disciplinary traditions need to learn at least to respect, if not to actually use, the methods and theories of the other. From my natural science side, we must recognise that just because we are working with people does not mean that we are 'doing' social science.

Then again, I am not advocating some perfect melding. Growth and improvement often seem to be driven from outside a discipline. It was pointed out by a peer reviewer of this chapter that fundamental incommensurability between conservation and some other disciplines may be important to acknowledge and possibly to embrace. Conflict and disagreement can be productive, and some have argued that the only reason conservation, and some of its practitioners such as myself, have moved in some of the directions indicated here is because of sustained criticism from the social sciences.

I am not suggesting some mash-up of disciplines but a more productive engagement—one that brings about improvement—even though that engagement may be criticism. Reyers et al. (2010) advocated broader adoption of transdisciplinary approaches in conservation, arguing that transdisciplinarity '... shifts the scientific process from a simple research process that provides a solution, to a social process that resolves problems through the participation and mutual learning of stakeholders' (Harorn et al. 2006, in Reyers et al. 2010). This sounds like a powerful way for social scientists to engage with the conservation community.

Geographers, anthropologists and political scientists have been attracted to conservation as a field of study, and more recently as a field of practise, and the social science literature on conservation has increased manyfold in the last 20 years. This is an important time for the discipline. Born out of the natural sciences, we have come to recognise that conservation can be effective only if it embraces perspectives, values and methods from a wide range of social sciences. This is not happening as well as it should or as rapidly as required. Neither social scientists with life sciences skills nor life scientists with social science skills are being groomed in any number. Unless this changes, we are likely to continue to suffer a split culture of action and critique rather than an adaptive practise informed by organised knowledge.

I have taken my hard-won appreciation for what the social sciences have to offer conservation back to my colleagues in the broader conservation community. There the social sciences are known and appreciated. Some social scientists have laboured long and hard within conservation organisations, although their expertise has not achieved the necessary change. But then, ways are changing as conservation careers are being pursued by people trained in the social sciences; even natural scientists are being exposed to the perspectives and approaches of social sciences. Additional changes will occur as some of the practising social scientists in conservation gain more leverage, as the Social Science Working Group of the Society for

Conservation Biology starts to spread its wings, and as more and more of us come to understand that our future success will rely on tools we can learn, borrow or modify from the social sciences. There is still a great deal more to learn.

So, this chapter is written out of frustration but also out of hope. With all the potential for social science and social scientists to reform and strengthen conservation, are not there ways to lift the veils obscuring the glass? Lack of progress is due partially to reluctance and neophobia from within the conservation community and a resistance to listen to careful critique. Nonetheless, I would like to suggest that there are many opportunities to improve this relationship that could originate with social scientists. This is not a new observation.

Peter Brosius pointed out to his fellow anthropologists that their discipline was more interested in identifying what is wrong with conservation than in trying to make things better (Brosius 2006). Similarly, social science colleagues have remarked to me that conservation is now considered a good field in which a young person can gain a reputation by adding to the literature detailing the crimes of conservation. Finding faults—although it can have its place—is easier than suggesting solutions. This serves only to make more difficult the productive engagement of the two areas. Nevertheless, progress has been made with thoughtful openings such as Peterson et al.'s (2008) list of 10 ways that conservation could be employed using a cultural lens perspective derived from anthropology.

The tools that have made the social sciences so perceptive, incisive and helpful to the human condition—the tools that characterise the book by Fairhead and Leach (1996)—are a means both of improving the practise of conservation and of sharpening social science's critique of conservation ideas and practises. With increasing consumption and demographic, climate and population changes over the next century, conservation will need an even closer engagement with people.

To meet this challenge we need careful and informed attention to the details of at least the following six conservation practises: (1) to variation, power, history, social constraints, and the influence of leaders and how their success depends on the vagaries of social and financial currents; (2) to the role of ego, competition, jealousy and market positioning; (3) to the difference between what is written and what is done; (4) to the power of the institutional *Eminence grise*; (5) to prejudice against one group and in favour of another; and (6) to chain migration, happenstance, persuasion and storytelling. These are some of the factors I have seen influence conservation practise.

We practitioners are not the monolithic conservationists depicted in much of the literature. Those concerned with conservation are humans in pursuit of human ends, with the foibles, powers and weaknesses, and institutional constraints that characterise all human endeavours. Rather than making hasty generalisations about the conservation community based on the writings of the few, and advancing simplistic arguments focussed on blame, not improvement, why not study us in our full natural history?

There is a lot of work to be done and a need for systematic analyses of a broad range of case studies, as well as a cessation of the use of case studies as good or bad indications of the worth of the whole. Learning about our practise and the structures that support it would improve our practise. Some, perhaps even most, in the conservation community stand ready. Nonetheless, some will continue to embrace the old ways and in that diversity of approaches will come strength.

In their conclusion Fairhead and Leach (1996: 3) emphasised the importance of recognising diverse interpretations of landscapes but warned that: 'Considering all landscape interpretations as in part socially constructed does not, however, negate the fact that certain readings can be demonstrated as false, and that historical evidence might support some more than others'. The themes of diversity of interpretations can mask the veracity of some interpretations over others. Some of what is being said by social scientists about conservation is absolutely true, but much of it seems not to be. Although many obstacles exist, there are openings for serious engagement by social scientists with conservationists and the broader conservation community. Researchers are working on institutional ethnographies and placing social scientists in the workplace of conservation organisations. There are probably other similar efforts with which I am not familiar. We need this work. We need to learn of, and from, our mistakes. We must improve our practise. For this, I maintain, we need the help—and informed criticism—of our social scientist colleagues.

Acknowledgments I would like to thank my social science friends and colleagues who have taught me so much but are not responsible for what I have learned: Bill Adams, Allyn Stearman, Katrina Brandon, Arun Agrawal, Bill Balee, Peter Brosius, Marcus Colchester, Steve Sanderson, Anna Tsing and Paige West. For perceptive and often critical, but always useful comments, I would like to thank Arun Agrawal, Nigel Dudley, Mike Mascia, Dan Miller, John Robinson, Steve Sanderson and Meredith Welch Devine. I also would like to thank the constructive engagement of three reviewers.

References

Agrawal, A., and C.C. Gibson. 1999. Enchantment and Disen Chantment: The Role of Covmmunity in Natural Resource Conservation. *World Development* 27: 629–649.

Brosius, J.P. 2006. Common Ground Between Anthropology and Conservation Biology. *Conservation Biology* 20: 683–685.

Chape, S., M. Spalding, and M. Jenkins, eds. 2008. *The World's Protected Areas: Status, Values and Prospects in the 21st Century*. Berkeley: University of California Press.

Dudley, N., ed. 2008. *Guidelines for Applying Protected Area Management Categories*. Gland: IUCN.

Dudley, N., J.D. Parrish, K.H. Redford, and S. Stolton. 2010. The Revised IUCN Protected Area Management Categories: The Debate and Ways Forward. *Oryx* 44: 485–490.

Fairhead, J., and M. Leach. 1996. *Misreading the African Landscape: Society and Ecology in a Forest-Savanna Mosaic*. Cambridge, UK: Cambridge University Press.

Harorn, H.G., Y.D. Bradle, C. Pohl, R. Stephan, and U. Wiesmann. 2006. Implications of Transdisciplinarity for Sustainability Research. *Ecological Economics* 60: 119–128.

Jenkins, C.N., and L. Joppa. 2009. Expansion of the Global Terrestrial Protected Area System. *Biological Conservation* 142: 2166–2174.

Mascia, M.B., J.P. Brosius, T.A. Dobson, B.C. Forbes, L. Horowitz, M.A. Mckean, and N.J. Turner. 2003. Conservation and the Social Sciences. *Conservation Biology* 17: 649–650.

Mcclanahan, T.R., J.E. Cinner, N.A.J. Grham, T.M. Daw, J. Maina, S.M. Stead, et al. 2009. Identifying Reefs of Hope and Hopeful Actions: Contextualizing Environmental, Ecological, and Social Parameters to Respond Effectively to Climate Change. *Conservation Biology* 23: 662–671.

Parks.it. 2017. Frequently asked questions about Italian protected areas. http://www.parks.it/indice/Efaq.aree.protette.html. Accessed 3 July 2017.

Peterson, R.B., D. Russll, P. West, and J.P. Brosius. 2008. Seeing (and Doing) Conservation Through Cultural Lenses. *Environmental Management* 45: 5–18.

Reyers, B., D.J. Roux, R.M. Cowling, A.E. Ginburg, J.L. Nel, and P. O'Farrell. 2010. Conservation Planning as a Transdisciplinary Process. *Conservation Biology* 24: 957–965.

Discussion Forum: Conservation Thinkers and Practitioners Respond

10. Steven R. Brechin: Returning Greater Integrity to the Conservation Mission in a Post-Neoliberal World

Neoliberalism is dead. The great global financial crisis of 2008 sealed its doom. Good riddance. I define 'neoliberalism' as a belief in the transformative power of free markets; that they can solve most problems and can self-regulate. It is paired with reducing the size and scope of government in managing the economy and society; public goods should be privatised. Most important to this commentary are all the forms of organising, from governmental agencies to nonprofit organisations; they should operate more like for-profit businesses to maximise their effectiveness. It is a business-centric model for organised action. It also rewards the accumulation of wealth and exaggerates the importance of those that have it, giving them a greater say in how almost everything is done. With the demise of neoliberalism as a global organising principle, the nature conservation mission, and its many organisations, must abandon neoliberal practise and return to the ideals of civil society space to restore greater integrity to their operations and to the conservation mission itself.

The 2008 financial crisis reminded everyone that government needs to regulate markets to protect the public interest. In addition, government must be a central player to adequately address rampant economic and social and/or racial inequality and ecological collapse. Over the decades, neoliberalism has led to the rise of a new superclass of ultrawealthy global elites, many of whom tend to have less of a stake in societal interdependence and democratic processes (Rothkopf 2008), requiring us to become more

P.B. Larsen, D. Brockington (eds.), *The Anthropology of Conservation NGOs*, Palgrave Studies in Anthropology of Sustainability, DOI 10.1007/978-3-319-60579-1

vigilant in protecting the collective good. In sum, we have learned that selfishness does not lead to collective welfare, but rather it frays the fabric of society and the human soul. As a philosophy, neoliberalism will live on, but we must endeavor to never let it dominate global society again. After 40 years of its existence as the major organising principle of almost everything, it has left deep wounds in our collective flesh that will take decades to heal.

The great Hungarian thinker Karl Polanyi, especially in his master work, *The Great Transformation*, warned us at the end of World War II of the corrosive influence neoliberal (laissez-faire) beliefs and policies would have on people, on communities, as well as the natural environment (Polanyi 1944, 2001). He feared the rise of a market society where social interactions would be profoundly shaped by the market's appeal to make everything into quantifiable commodities affecting even human morality (see also Sandel 2012).

Polanyi warned as well that implementing a global, self-regulating free market system, based on the Libertarian principles espoused by Ludwig von Mises and Friedrich A. Hayek—the core elements of modern conservativism—would bring more chaos to the world, not less. Nevertheless, Polanyi did not believe that markets themselves were evil. He saw them as essential for guiding economic activity and providing for people's needs. But markets require guidance by institutions that reflect the common good, not simply their own amoral logic. To Polanyi, authoritarianism did not only come from the left, per von Mises and Hayek, but also from the right.

Although neoliberalism is dead, it needs to be buried and very deeply, with a stake in its heart, so that it cannot rise again. To fully put it to rest, however, people must first know that it existed. Even many academicians, especially those outside of anthropology, geography, political science and sociology, are uncertain what it is and have a hard time linking it to real-world consequences. The public has no clue, at least in the United States. So, in every undergraduate class this writer teaches, I ask my students in the very first lecture: Which political ideology has shaped your lives and those of your family, the nation, and the world over the last 40 years? They look at me like I am from another planet.

Most do not know neoliberalism by name, or how it has affected them, although they are upset about growing social and economic inequality, a democratic political system more responsive to the wealthy few than to the majority, and especially about their college debt burdens because of the commodification of higher education (Wong 2016). But when I ask them what can be done to foster social change, they typically respond, after a long silence, with suggestions about purchasing choices, what we might buy and from whom—all market-based. Fortunately, this is beginning to change. There is now growing social and political activism. In the United States at least, with the very recent rise of a progressive left represented by political leaders (e.g., Bernie Sanders and Elisabeth Warren) and activist groups (e.g., Occupy Wall Street and Black Lives Matters), there seems to be a growing acknowledgment that a new organising ideology is required.

Still, the activism is neither named nor contrasted with neoliberalism. But then, there is the British Brexit vote and the rise of Donald Trump. Both events come out of misplaced anger by those that neoliberalism-based globalisation left behind. It is an anger that should be more forcefully directed largely to that agenda. Ordinary people are angry and they should be, but they incorrectly blame basically neutered governments and redirected focus on 'the other' (e.g., immigrants, for their plight), not on our foundational ideology of the past 40 years. A missed opportunity.

We should not make the same mistake when it comes to conservation organisations. Most are nonprofits, as well as strange hybrids for example, the International Union for Conservation of Nature and Natural Resources (IUCN), partly a nongovernmental organisation (NGO) and partly an international governmental organisation (IGO). Many of them eagerly boarded the neoliberalism train. One major conservation group, Conservation International (CI), was founded very much on its principles. At one level, it was not difficult to blame them. Having membership-based engagement in the its mission reduces management flexibility and organisational agility. More funding, regardless of the sources, meant more resources for conservation initiatives—a higher purpose.

In addition, may there be no doubt about the vicious competition among conservation organisations for the attention of dollars, especially from wealthy individuals and corporations. Fundraising has always been challenging. Nevertheless, growing the organisation was rewarded, as it is in business, with greater organisational stability, higher CEO salaries, celebrity events and very questionable partnerships (McDonald 2008; Brockington 2009). This hypercompetition for dollars surfaced immediately when I worked for a short time behind the scenes on the horrific politics following the publication of Mac Chapin article's in *World Watch Magazine* over a decade ago (Chapin 2004). The World Watch Institute received threats from the other conservation CEOs that they would go after World Watch's well-healed donors if they would not put to rest the (in)famous article's controversy immediately. How is that for higher purpose? Criticism could not be tolerated in fear of threatening donations.

During the era of neoliberalism, conservation NGOs put organisation-building ahead of the mission. In fact, the two became linked. It was framed the same as that as the organisation and its budgets grew the more 'good work' it could do. This may sound logical at face value, but presents a slippery slope where many slid too far. I encourage conservation NGOs to work with large corporations to make them or their practises more ecologically thoughtful. This is essential work. What I do object to is their accepting money from them to do so. This is called a 'conflict of interest', a forgotten concept that needs to be reintroduced.

Likewise, conservation organisations should not give corporations, especially those harming the planet, major conservation awards because they gave a large donation to the organisation. This is called 'bribery', at worst, or 'quid pro quo' at the very least. In the business world, this is just doing business. Such actions by conservation organisations, however, could only be justified if organisation-building, as it was often suggested, is the same as the conservation mission itself. They are not, of course. To be frank, all of this does not wash down well with the public and other critical conservation stakeholders.

One of the most stunning interviews I did while exploring large conservation NGOs several years ago was when an eager employee from a corporate engagement at The Nature Conservancy (TNC) headquarters declared, with enthusiasm, that 'we take bad money and turn it into good money' as if conservation NGOs were money-laundering operations. This is 'greenwashing' at its most egregious, but it not only was framed as normal but also as desirable. The trouble with such thinking is that it ignores the notion of integrity and the need to maintain the public's trust.

The public needs to know the organisations that occupy its civil society space (given their tax-exempt status) have integrity, where the *means* are just as justifiable as the *ends*. We expect civil society organisations, our nonprofits, to follow higher moral principles, and that they not be corrupted by large, wealthy actors of any stripe. This does not mean they should not engage these actors to solve the world's problems. They must! But these relationships need to be maintained at arm's length to assure credibility. Wealthy individuals passionate about the conservation mission should be welcomed onto conservation boards if they have expertise to bring to the organisation, but not as active corporate CEOs.

Space does not allow acknowledging the full nuances of raising revenue and working with other important large actors. Nonetheless, changes must be made. I am aware that these suggestions may impact the budgets, staffing and salaries within conservation organisations. Yet, rebuilding integrity into the conservation mission will require conservation NGOs to rethink much of what they do and how to rebuild their funding streams. Still, integrity is essential in a reinvigorate civil society space in a post-neoliberal world (Edwards 2010) because it will draw more people to the cause.

Major funding organisations and foundations must increase their share of funding for these nonprofits to replace the revenues loss from corporate giving. This along with broad public outreach are essential to the future of conservation. Ideally, nonprofit conservation organisations must be more closely tied to the public, or at least its membership if that membership is broad. In sum, conservation organisations must return fully to civil society as a 'citizen space' where groups act more for the collective needs of society than for the organisation itself, or as the public relations wings of the corporate community. Of course, The Nature Conservancy is built on its membership. Thus, one needs to acknowledge that broad membership does not prevent all missteps, but it helps. Conservation International now accepts small donations from ordinary citizens. Earlier they scoffed at the idea. This is a step in the right direction.

Some will argue (and have) that ends justify the means. That conservation goals need to be achieved regardless of how. The problem with that argument is that it taints the larger conservation mission and makes the public and other stakeholders cynical, if not outraged by, the process. Without wide public and stakeholder support the overall conservation mission will falter over the long term. The public hold nonprofits and other civil society groups to higher standards than government and certainly corporations. Nonprofits are seen as having 'purer souls', as organisations of the people, even though this view is not always justified. Still it is this greater public legitimacy that is precisely why corporations want to partner with them, as well as precisely what is at stake.

Conservation should be about protecting nature and the human community depending on it, not the pursuit of profits to grow the organisation. In organisational theory this substituting one grander goal with a narrower, more immediate one, is called 'goal displacement' (Warner and Havens 1968). It has been going on far too long in the conservation community. Growing conservation organisations by any means necessary does not enhance the conservation mission, it diminishes it. Nonprofit conservation organisations must fiercely protect their integrity and the public's trust. If not, the value of the conservation mission itself will become compromised. When that happens, we all lose.

References

Brockington, Daniel. 2009. *Celebrity and the Environment: Fame, Wealth and Power in Conservation.* London: Zed.

Chsapin, Mac. 2004a. A Challenge to Conservationists: Can we protect natural habitats without abusing the people who live in them? *World Watch Magazine*, 17 (6), November/December.

Edwards, Michel. 2010. *Small Change: Why Business Won't Save the World.* San Francisco: Berrett-Koehler Publishers.

McDonald, Christine C. 2008. *Green, Inc. An Environmental Insider Reveals How a Good Cause Has Gone Bad.* New York: Lyons Press.

Polanyi, Karl. 1944/2001. *The Great Transformation: The Political and Economic Origins of Our Time.* Boston: Beacon Press.

Rothkopf, David. 2008. *Superclass: The Global Power Elite and the World They Are Making.* New York: Farrar, Straus and Giroux.

Sandel, Michael J. 2012. *What Money Can't Buy: The Moral Limits of Markets.* New York: Farrar, Straus and Giroux.

Warner, W. Keith, and A. Eugene Havens. 1968. Goal Displacement and the Intangibility of Organisational Goals. *Administrative Science Quarterly* 12 (4): 539–555.

Wong, Alia. 2016. The Commodification of Higher Education. *The Atlantic*, March 30. http://www.theatlantic.com/education/archive/2016/03/the-commodification-of-higher-education/475947/.

Steven R. Brechin is Professor of Sociology at Rutgers University.

11. David Cleary: What are the Grounds Needed for Dialogue?

As a card-carrying, dues-paying member of the 'transnational conservation elite',[1] having done my part in transforming 'modernist biodiversity conservation [into] an organised political project',[2] not to mention repeatedly 're-enacting NGO identity narratives',[3] even though at the time I thought I was helping to create protected areas (PAs) or strengthening indigenous organisations, it was a fascinating exercise to read the contributions to this book. This writer believes there is potential for a fruitful and mutually enlightening dialogue between the academy and conservation practitioners, an assertion I ground in my own background: doctorate in anthropology and young academic at the universities of Edinburgh, Cambridge and Harvard That is before falling by accident into conservation, not through a sense of vocation but the instability of career paths at junior levels of academies and a wish to live in Brazil. To my surprise, and certainly not as part of a plan to make a career in conservation, I enjoyed the work and ended up staying.

At that time—I joined TNC in 1999—Redford's description of conservation organisations as dominated by biologists and conservation biology was accurate.[4] Times have changed. Beyond the Wildlife Conservation Society (WCS), it is now rare to find biologists in very senior management positions in the global environmental organisations. In my peer network, there are more social than natural scientists; some anthropologists and geographers, but more economists. The single largest group, a topic to

P.B. Larsen, D. Brockington (eds.), *The Anthropology of Conservation NGOs*, Palgrave Studies in Anthropology of Sustainability, DOI 10.1007/978-3-319-60579-1

which I will return, is one with no particular academic background at all, but experience working in the private sector. There is also a depressingly large number of lawyers.

You can leave the church, but the church never leaves you. A concern with representation and ways of thinking, being aware of culture wrapping itself comfortably around any formal structure, thinking oneself into other people's shoes—all these anthropologist habits are useful in organisational life. An unexpected consequence of leaving the academy was freedom to read more broadly; from a long list of potential citations I would single out Simon Schama's *Landscape and Memory*, Hugh Raffles's *In Amazonia* and James Clifford's *Person and Myth* as books I have returned to again and again when thinking about the practise of conservation. More specifically, thinking about the social construction of nature and landscape, and how to navigate between popular and historically informed definitions of what and how to 'conserve'. More recently, Isaiah Berlin's exploration of the pursuit of incommensurable goals as a defining element of the human condition has helped me relax a least a little as I reflect on how conservation organisations and those within them go about their work. There would be much to be said in a serious dialogue between environmental organisations and the academy.

I take this autobiographical approach since it leads into the source of my disappointment with this book. Anthropology is about many things, but one is trying to understand how agency works, or fails to work, or starts to work and then lopes off in unexpected directions. Nuance and a grasp of history are essential to it. From that standpoint, I found little to engage with in this book's chapters. From an ethnographic perspective, it is unsurprising that Redford's is in some way the strongest chapter, given his insider status at WCS. Larsen's dealings with WWF also inform his more convincing, albeit still incomplete, account of NGO dynamics—although choosing 'Dirty Harry' when the obvious film to reference was *The Good, The Bad and The Ugly* was what struck me most about his chapter. Even though the case studies presented by Nuesiri in Chap. 8 and Ruysschaertt and Salles in Chap. 5 were interesting as far as they went, they were case studies. On a higher level, little else engaged me, unless frustration and disagreement count as engagement. So, what is the problem here? What did most contributors miss, and why did they miss it?

I believe the issue is a conceptual framework grounded in political economy, and to a lesser extent in discourse analysis, that is inimical to agency and rather ahistorical. It runs through all except Redford's chapter, to differing but often overwhelming extents. It left me thoroughly depressed about the possibility of substantive dialogue with the academy because the nature of the framework raises the question of why it would be worth my or any other practitioner's time to engage with, given that it seems to me to foreclose a real exchange of views, exclude centrally important types of information, and lead those committed to it to fundamental misunderstandings of how conservation organisations work.

Too many of the chapters are crippled by a grounding in a neo-Marxist political economy, symbolised by the recurring use of the term 'neoliberal'. As a general, very high-level term, it has some descriptive value as shorthand for the ceding of certain spaces and forms of provision by the state to the private sphere. But that descriptive value is limited, given that shift has been everywhere incomplete, even in the 'neoliberal' heartlands of the US and the UK, where government spending as a proportion of GDP has risen, unevenly but constantly, since 1980. It is the differences within the very broad shift toward neoliberalism, both geographical and in terms of institutional approaches, that make it a loaded and dangerous analytical term when used by contributors such as Holmes, MacDonald, Blanchard and even Larsen. Anyone attuned to the varieties of historical experience between countries over the last 30 years, even between neighbouring countries featured in different chapters here, such as Peru and Chile, is alienated by a conceptual apparatus that operates at this level of flattening generality. Yet, to say it does not do nuance is only half of the critique.

Because the reality is that 'neoliberal' is used not only as a descriptive term in these essays but also as an ideological one, or rather more than ideological, as a puritanical expression of disdain that has a long tradition in some academic circles. Any engagement with markets, any attempt to influence the impacts of markets on the environment, any addressing of the relationship between markets and land use and/or biodiversity becomes 'neoliberal' thinking or a 'neoliberal' approach—a term that shuts down a conversation rather than opens it up. This is because many practitioners, myself included, simply do not accept, and even find rather offensive, the implicit identification of our political beliefs and working lives with Hayek and That Bloody Woman. This may come across as a strong statement, but what other construction is there to put on such an assertion: '[I]t is necessary to view modernist biodiversity conservation as an organized political project'[5]; or the notion that organisations like my own represent 'a new organizational order in which the interests of capital accumulation receive an unparalleled degree of access and consideration in conservation planning and practice'[6]?

Ideological disagreement I can take, but when conceptual rigidity leads to basic empirical mistakes, it is less easy to forgive. This is a framework where the most contentious statements are presented as self-evident; in a minor key, Holmes's assertion that modern Chilean history can be summarised by saying 'there have been only extremely modest retreats from the principles and approaches of the Pinochet era reforms' (!!), or, in a major key, the idea repeated by several authors that NGOs are not really organisations at all but a loose collection of projects run by a single fund-seeking agent, or the notion that movement of staff between organisations means that NGOs are somehow unstable—all that is solid melting into air.

Really? Communities of practise exist everywhere, and the fact that software engineers or academics regularly switch jobs does not make Google or Microsoft or Oxford or Yale institutionally unstable—rather the opposite. The larger transnational NGOs are far more than a collection of projects; they spend more effort and resources on central functions (e.g., building science capacity) to give more weight to interventions in public debates in areas, such as climate change, lobbying and advocacy. Their aim is to steer both public and private resources toward conservation, consciously working as much as possible at a system rather than a project level. This is distinct from fundraising in the narrow, project-level sense, which is part of the portfolio of activities big international NGOs (BINGOs) engage in, but for several of the most important—World Resources Institute, Greenpeace—a relatively minor part. This higher level, more strategic work is missed by all the contributors save, once again, Redford.

Therefore the naive nature of Brockington's assertion that, in sub-Saharan Africa 'only 14% by area of all protected areas received any form of support from conservation NGOs'. What is meant is that staff from NGOs can only be found in the field at 14% of PAs by area, but that ignores the much more important work NGOs do on finding resources for protected area systems as a whole. In general, I would argue that even more important to 'modernist' conservation than an engagement with markets has been the expansion of PA systems across the globe and the intense engagement with governments and public policy that underlies it, which sits rather uneasily with an insistence on 'neoliberal' approaches as central to what we do.

Yet the contributors are right in arguing that modern conservation thinks much more about markets, supply chains, market-led approaches and engaging the private sector than it did when I joined the Conservancy. I mentioned previously that the pace of recruitment of people with private sector experience to senior positions has picked up in recent years. Partially, this is because NGOs, like companies, require professional management, but it is also part of a trend to engage more deeply with markets and market mechanisms.

Still it seems in this book that *any* engagement with markets and market forces is characterised as 'neoliberal' and implicitly collaborationist—or not so implicitly by MacDonald. This reductivism obscures the reality that NGOs do not run only one set of strategies but several: market and non-market, public and private strategies are run simultaneously *and would not work otherwise*. I used the word 'naivety' earlier to describe this lack of ethnographic literacy, but it is conditioned by a binomial political (and moral) economy incapable of dealing with the fluid, shifting boundaries between the public and the private that characterises modernity. A framework that labels any market engagement 'neoliberal' and assumes public–private shifts only operate in one direction cannot cope with complexity. No reader of Holmes's chapter, for example, would know that the largest private land purchases are an intermediate stage in transitioning that land into public ownership in Chile (and Argentina!). Several national parks in Patagonia today began life as private land purchases by Tompkins and others, a central plot element that somehow got lost in the telling.

Which leads to another problem. Not once did any contributor—as usual, excepting Redford—cite a success or achievement linked to engaging the private sector or market mechanisms. The alleged lack of success validates the critique presented. Yet such success abounds. On a macro-level, how else could one describe the steep fall (around 70% in a decade) of deforestation in the Brazilian Amazon, partially because of targeted and effective regulation and civil society pressure, but also to voluntary private sector agreements in the beef and soy industries? How else can one describe the commitments given by dozens of companies to *eliminate* deforestation, not reduce it, from soy, beef, pulp and paper, and palm oil supply chains by 2020? Commitments given not in unreliable industry spaces subject to greenwashing but formalised in the UN's 2014 New York Declaration on Forests and Climate, a political commitment where failure to implement will have major political costs for the many companies concerned.

On a landscape level, how else can one describe the expansion of social and environmental certification in smallholder-dominated landscapes producing cocoa in Ghana, coffee in Guatemala, tea in Sri Lanka, and many other commodities in many other countries? On a micro-level, how else to describe increases in avian species on forest/field boundaries, increases in pollinator species given certain types of land management and crop rotation regimes, increases in butterfly diversity in reduced impact logging systems? All these are a direct result of market and supply chain interventions, working directly with companies. There is a vast grey and formal literature documenting all these levels of success which is almost religiously ignored in this book.

This blindness is not accidental: it is an induced, conditioned blindness that belongs more, I think, to a MacDonaldian ideological project than the openness and attention to the empirical that I take as central to academic enquiry. If there is to be real and fruitful exchange between the academy and NGOs, which several of the authors claim to want, I suggest a rather more flexible and nuanced conceptual approach. I was mystified, for example, that the political economy used by the authors emphasised neoliberalism rather than globalisation. A proper attention to globalisation would have enabled a much subtler approach not just to institutional practise but also to discourse and narrative—to forms of social construction of 'environment', 'nature' and 'conservation'—as well as being more consistent with the transnational nature of modern conservation.

Rather than seeing power relations through a Manichean lens of dominance and submission, hegemony and subordination, it would allow for a properly textured rendering of representation and politics in developed and emerging economies. It would allow one to start to render the complexities of a country, such as Brazil, to take the example I know best, where the technology of tracking land use in tropical ecosystems is the most advanced in the world—as is the regulatory framework for agriculture. There too highly sophisticated national and local NGOs are increasingly influential in transnational spaces as well as within their own national and local boundaries. If there is to be dialogue between the academy and conservation NGOs, the academy needs to do some self-reflection and move beyond the cruder kind of political economy too often displayed here toward a more historically informed conceptual framework capable of capturing the empirical, discursive and political complexity of transnational conservation in a globalising world. Reductive approaches are not conducive to dialogue.

Notes

1. Blanchard et al., see Chap. 6 of this book.
2. MacDonald, see Chap. 4 of this book.
3. Larsen, see Chap. 2 of this book.
4. Redford, see Chap. 9 of this book.
5. MacDonald, see Chap. 4 of this book.
6. Idem.

David Cleary is Director of Agriculture at the Nature Conservancy.

12. Ashish Kothari: Rethinking Conservation Epistemologies

This commentary is based mostly on a reading of Ken Redford's paper and a glance through Peter Larsen's, but may have relevance also to some of the other chapters too, judging by their abstracts (which is all that I have read of them).

As always, Redford is stimulating; it is refreshing to see his evolution with an increasing acceptance and understanding of the social, political and economic dimensions of conservation. Not all conservation scientists have been willing or open enough to take this journey; and, in agreement with what Kent says, not all social scientists too have been willing to have themselves challenged regarding ecological issues. I concur with his contention that both the natural and social scientist, to optimise what they can offer each other, must shed further stereotypes and be open to collaborations that can make conservation both more effective as also more democratic and socially sensitive. There are some areas of emphasis that I may differ with Kent on, and I am tempted to challenge him to 'prove' (since he is a scientist!) that 'some of what is said by social scientists about conservation is absolutely true but much of it seems not to be'. (Surely he does not even have access to most of what social science is writing about conservation, given that much of is in languages he does not comprehend or in grey literature not easily available!) But then again, these are not the main points of my comment.

P.B. Larsen, D. Brockington (eds.), *The Anthropology of Conservation NGOs*, Palgrave Studies in Anthropology of Sustainability, DOI 10.1007/978-3-319-60579-1

Where I think Kent and many other conservationists (and, equally, social scientists) need to go much further is in rethinking epistemologies based on much closer, collaborative work with indigenous people and other traditional local communities. It is necessary to collaborate with each other as scientists (assuming that by this term is meant 'those who work in modern disciplines'), but it equally is necessary to work with those who deal with the rest of nature through other knowledge systems. There is increasing attention to the 'epistemologies of the South',[1] with roots, logic, and ways of understanding and expression that can be radically different from 'Western' or 'northern' (modern) epistemologies. This is emerging in a fledgling way in several academic places around the world, but much wider is its articulation by indigenous people and other traditional communities as part of their struggles to resist colonisation by nation-states, or by the industrial economy, or as part of their assertion of being autonomous stewards of the land.

People who live amid the natural landscapes and near seascapes that conservationists want to protect have their own ways of understanding them. For a long time now they have pointed out that the biological and the cultural are not necessarily neatly divided (thus the move in some parts of the world to have biocultural landscape recognised), just as the 'human' and the 'natural' are not. Worldviews that encompass the spiritual, the material, the cultural, the biological and much else, in an integral whole, are difficult for the compartmentalised sciences to comprehend and appreciate, but scientists have to make the effort (as indeed some are).

This is also about a clash of civilisations. When the UK-based Vedanta Resources, a London-based mining company came looking for a lease in the Niyamgiri Hills of Odisha in southeast India, the Adivasi (indigenous) Dongria Kondh said the land and forests belong to Niyamraja, their supreme deity, and no one has permission to 'strip his land'. Offers of high compensation, of jobs, of 'development' did not convince the Dongria Kondh. Some civil society organisations supported them through scientific studies that showed the conservation importance of the Niyamgiri Hills. This and collaboration in taking the matter to the courts was helpful in eventually rejecting the mining proposal. Nonetheless, through all this there was little attempt at understanding the worldview of the Dongria Kondh, in their own terms, and presenting this as an alternative to the dominant 'development' mindset pushed by the government and corporations, and even by many civil society organisations and scientists. So, no one noticed an even greater (albeit more long-term) threat to the Dongria Kondh, the welfarist state that was bringing completely inappropriate education and developmental efforts into the area in the name of civilising the tribals. Such a move that over time would destroy the identity and autonomy of these people, and open their landscapes to commercial pressures through their own young people who would have been 'educated' into thinking that industrialisation is the best path toward progress.

I give this example in some detail because it illustrates the point I am making. Had the collaboration between scientists, conservation and human rights groups and the Dongria Kondh been deep-seated, helping to understand different knowledge systems and ways of knowing, the subtle dangers to the Adivasis and to the biodiversity of their hills would have emerged much earlier. Fortunately, it is now coming out, and hopefully it is not too late.[2] I suspect that this would be the case in many parts of the world. Conservation groups would benefit greatly from breaking out of their epistemological shells and understanding those of the people who live amid the nature they want to conserve, through methods of mutual learning and appreciation. They can then be much stronger allies in the struggle against the juggernaut of globalised 'development', which will otherwise undermine the constituencies of both the natural scientist (nature) and of the social scientist (communities).

I do not want to underestimate the difficulty of this exercise. Each of us is immersed in a distinct upbringing that provides us the tools to understand the external world, and learning new tools, especially if they can appear to challenge the existing ones, is difficult. The first step is to simply acknowledge that there are knowledge systems and ways of knowing that are different from ours, a step that many scientists (and conservation groups of the global North[3]) have a difficult time taking. There are exceptions, of course, and this would be an added dimension to the point that Larsen makes about more nuanced understanding of conservation NGOs. If one does not have the time to immerse oneself adequately to also learn other ways of knowing, at the least with this first step of acknowledgment and recognition, one would have learned how to respect them. This is as (or perhaps even more) crucial to achieving conservation success and democratising conservation, as is a closer collaboration between natural and social scientists.

Notes

1. See, for instance, the work of Boaventura de Sousa Santos, and/or Maori women scholars (e.g., Linda Tuhiwai Te Rina Smith).
2. See Tatpati, M., A. Kothari and R. Mishra. (2016) *The Niyamgiri Story: Challenging the Idea of Growth Without Limits,* Kalpavriksh. (http://www.kalpavriksh.org/images/alternatives/CaseStudies/NiyamgiricasestudyJuly2016.pdf)
3. Which includes many conservation groups working in the South, the members of which have imbibed the formal conservation attitudes that have dominated the last few decades.

Ashish Kothari is a Founder-Member of the Environmental Group Kalvapriksh.

13. Diane Russell: Beyond NGOs: The Institutional Imperative

Many chapters in this book raise questions about the nature of conservation NGOs and the wider institutional setting within which they work. This commentary[1] focuses on institutional arrangements broadly rather than one type of institution, conservation NGOs. It seeks to reveal systemic factors and patterns that can generate resonant solutions for conservation.

All actors in conservation work within institutional settings, whether these be local groups or administrative units, research and educational institutions, donors, government agencies or NGOs. Every institution has its imperatives, incentive structures, systems and (overt and covert) power relations. When institutions interact, these elements come into play. Individuals can form bonds across institutions to achieve mutual objectives, and projects can be designed to foster collaboration among institutions; nonetheless, these efforts are framed within the institutional structures in place.

Like all institutions, conservation NGOs are diverse and rapidly evolving. Within the organisation there are (overt and covert) leaders, followers and dissenters, and these roles shift depending on topic or process (as Blanchard and colleagues hint at in this book). As we have learned, discourse shapes and is shaped by these social/power relationships. Forces inside and outside of the institution influence the discourse and there is typically a flow between what leaders want or feel they need to sustain the institution, or their part of it, and which outside influences get to be heard. A rights-based conservation discourse will attract a certain group of

P.B. Larsen, D. Brockington (eds.), *The Anthropology of Conservation NGOs*, Palgrave Studies in Anthropology of Sustainability, DOI 10.1007/978-3-319-60579-1

people and some NGOs or think-tanks will capture this discourse and hang their hats on it. Outside influences that reinforce that discourse will be privileged. A 'leveraging the private sector' discourse, taken up by several chapters, will capture others. This discourse is especially compelling because the leveraging is thought to be multifaceted: magnifying influence, scale of potential impact and financial resources.

Donors may attentively listen to conservation NGOs, and fund their work, or not, depending on the dominant or emerging discourse and, of course, the imperatives of their own funding streams. When funding diminishes or ends, an NGO or other institution dependent on external funding will try to adapt through adding or reducing staff, diversifying funding streams or other measures. They also try to (re)shape the discourse to favour their approaches.

On the 'recipient' end, the will to cooperate and the ability to produce measurable gains for conservation and people emerges from the benefit stream—its appropriateness, magnitude, diversity and sustainability. Benefits can be financial, social, political or intellectual. Patron–client links (e.g., NGOs train elites) are one form of benefit stream that can appear to keep things moving but will not bring enduring impact. Some NGO efforts incorporate this understanding, or grow into it. They seek multiple funding streams and bring technical skills to ensure appropriateness, scale, endurance and diversity. Many do not. But then again, many other institutional actors—researchers, government agencies, donors—also do not have the requisite experience, resources or endurance for effective and just conservation action. Ruysschaert and Salles, Chap. 5 in this book, illustrate that sort of diversity.

There is also the fact that conservation outcomes are firmly shaped by factors far removed from a site or proximate threat. The NGOs get blamed for bad outcomes in situations where autocratic governments have ruled for generations (e.g., Central Africa) but get credit when outcomes are good in a country committed to good governance (e.g., Namibia). Failure to work across sectors and levels to address deep drivers of biodiversity loss falls across all parties in conservation. The nonprofit, research and donor worlds are increasingly fragmented and specialised, even internally. When we deploy 'big picture' and 'systems thinking', we can make headway in analysis; however, it is extremely difficult to mobilise the resources to translate this analysis into long-term comprehensive programs. Brockington and colleagues' 'thought experiment' described in this book will remain just that.

So, we have to ask the question: '[D]oes it really matter who implements a conservation initiative?' Maybe not. Maybe what matters most is *how* it is implemented. But do we know what makes for a good approach?

Learning and sharing knowledge about approaches and outcomes is hampered by institutional imperatives. Institutions need to feel ownership over approaches and concepts that they can market—whether that be to donors, to prestigious journals or to local communities. Being attentive to peers and leaders within one's own institution and 'community of practise' is part of career survival, and this often means that people have little time or incentive to learn about other ways of thinking.

To complicate the situation, evidence for effectiveness and equity in conservation is contested and sparse. There is institutional resistance to and confusion over what constitutes evidence. What is considered solid evidence in one institutional setting may be rejected or strongly questioned by another. Evidence that reveals negative outcomes can be felt as a threat to institutional survival. A good example is the dueling evidence for the effectiveness of indigenous forest management in reducing deforestation.

More fundamentally, various institutions speak different languages: lexicons, tropes and memes embedded in institutional discourse can make cross-institutional communication difficult. In addition, at an even deeper level, individual and institutional epistemologies may differ radically. Some actors see conservation through the lens of threat, while others look through the lens of opportunity. Despite decades of trying to integrate conservation and human rights (i.e., social development), differing epistemologies persist and they shape actions and funding streams.

Institutional arrangements and imperatives are not going to disappear. We, however, can make them more legible as this book seeks to do. We also can nurture 'culture brokers' who cross institutional and epistemological boundaries and foster genuine collaboration. Within institutions we can identify assumptions embedded in narratives that need testing and push to test them. We can reduce jargon, whether it be bureaucratic or academic, to enhance communication.

Finally, we can participate in analysis and fieldwork with an array of actors. We can test our assumptions and build an evidence base together that incorporates an institutional context. This is what allows us to craft genuine partnerships while also being alert to institutional arrangements and relations of power and privilege.

Notes

1. Diane Russell's views are her own and do not represent the views of the USAID or the United States government.

Diane Russell is an Independent Consultant with Experience in the Donor Community.

14. Kartik Shanker, Siddhartha Krishnan, and Marianne Manuel: The Politics of Conservation: Deconstructing The South

It has been proposed that the globalisation of conservation has been affected by large international conservation organisations (INGOs or BINGOs) that, somewhat analogous to the spread of capitalism by multinational companies (MNCs), have spread northern and/or Western approaches to conservation across the world, sometimes to the detriment of both environment and local livelihoods in developing countries (Rodriguez et al. 2007; Chapin 2004).

As Chap. 3 by Brockington et al. in this book points out, there is reasonable evidence to support this view. However, this is not the entire story. The conservation NGO sector is remarkably complex, and there is considerable politics that defines the identity and actions of actors on local, national and international scales. We provide evidence for this using two examples from India, one terrestrial and another marine, that speak to the complexity both within and across scales.

Two Turtle Tales

Sea turtle conservation is indeed a global enterprise, following on the heels of large charismatic mammals in terms of attention and funding, and perhaps even greater in its spread. From small local to large international organisations, there are a range of philosophies of and approaches to the practise of sea turtle conservation. Here, we provide two contrasting cases from Odisha, a well-known breeding site on the east coast of India.

P.B. Larsen, D. Brockington (eds.), *The Anthropology of Conservation NGOs*, Palgrave Studies in Anthropology of Sustainability, DOI 10.1007/978-3-319-60579-1

More than a 100,000 turtles nest during unique mass nesting events (called 'arribadas') at two sites in Odisha, the largest rookery outside Central America. However, the conservation of olive ridley turtles poses significant challenges. Tens of thousands of olive ridley turtles are killed as incidental catch in trawl fishing nets each year; although there are fisheries laws that prohibit nearshore mechanised fishing, these are not properly enforced. Traditional fishing is allowed in most areas (except during the nesting season), and in fact the original fisheries laws were instituted to protect them from the mechanised sector. Nevertheless, the promotion of these laws as turtle conservation measures and the demonisation of trawlers has led to a bitter conflict between fishing communities and the state as well as conservationists (Shanker and Kutty 2005).

On the other hand, the traditional sector varies across the entire coast, and includes a variety of gear and craft types that range from entirely manual to significantly mechanised. In 2003 and 2004, the Government of India issued notifications that prohibited fishing by the traditional fishing communities in certain turtle congregation areas. In their response, traditional fishing communities argued that their methods posed little harm to turtles and that they should be considered as partners in turtle conservation. Ostensibly in support of this, they locally banned three types of nets that they claimed were harmful to turtles. Still, a closer examination of this remarkable measure reveals that these nets were used by those fishing for a living from a different community in the adjacent state—the southern Odisha fishermen were attempting to keep them out. A few years later, the traditional fishemen had become a part of a network, the Orissa Marine Resources Conservation Consortium, as they clearly saw the benefits of using conservation policy and NGO support to further their agenda at the local level.

Over the last decade, conservation itself has flourished as a livelihood through the formation of local NGOs, in a similar manner as in the case of conservation in the Amazon illustrated by Larsen in Chap. 2. In fact, recent surveys in Odisha indicate that, as alternate livelihoods, they see more potential for funds and/or jobs in conservation than in ecotourism (Ridhi Chandarana et al., unpublished data). Clearly, communities were able to recognise that they could tap into national and global conservation movements to further their own cause, in one instance as a community action, and in the other(s) by recognising conservation as an opportunity for income and upward social movement.

At the other end of the power spectrum, a fascinating struggle developed when several large national and international NGOs grappled with the development of a port near one of the mass nesting beaches in Odisha

(Shanker 2015). When the Indian multinational conglomerate, Tata Group, joined a consortium to build a port in Dhamra, they were interested in commissioning a study to ascertain whether the port would affect olive ridley turtles. The Tata stated that it was concerned about the conservation of ridleys and would either withdraw from the project if it impacted turtles or would take suitable mitigatory actions. During a campaign against the port that involved a variety of individuals (biologists, conservationists) and organisations (from local to national to international NGOs), every conceivable alliance and adversarial combination occurred.

Many Indian NGOs declined to carry out or protested the study, as some believed that construction should stop while the studies were ongoing, while others believed that any engagement would lead to greenwashing the port. These included both local NGOs as well as Indian counterparts such as Greenpeace and WWF. At this stage, the Tata invited the IUCN, which in turn requested the Marine Turtle Specialist Group (MTSG) to take on the task of advising the company. Ignoring the protests of the local MTSG experts as well as of IUCN member organisations, MTSG/IUCN carried out a scoping study on the impacts of the port and advised on the impact on mitigating turtles of dredging and lighting.

The three large international NGOs had three completely different approaches. Greenpeace was going to campaign against the port (and any individuals who might work with the company) regardless of what other opinions were offered. The IUCN had decided to work with the Tata no matter what any local expert or organisation said, displaying a complete lack of respect for local opinion. Both, though they took exactly opposite positions regarding working with corporations, were similar in that they were going to follow a path that had been determined in advance. In this instance, WWF took a relatively balanced approach and acted with the majority.

Over a two-year period, a series of strategic alignments were formed, many of which included environmental NGOs aligning with the corporation, and several that included NGOs battling each other. What local as well as international NGOs chose to do was driven by a combination of politics, ideology and 'business interest', rather than a clear Western–international versus local dichotomy.

Fueling the Forest Fire

Redford, in Chap. 9 in this book, rightly points out that conservation is not solely the domain of the Western elite but is often a product of local culture and context. While agreeing with this statement and the need to resist imposing Western ideas of conservation onto groups in developing nations, we must point out that the idea of what conservation means for these two groups is not always vastly different. For instance, one of the tenets of the 'Western approach' to conservation is the establishment of a pristine wildernesses for conservation. Inevitably, this has involved the displacement of local communities and indigenous people. Large NGOs have been squarely blamed for the creation of such conservation refugees (Dowie 2005; Chapin 2004). Without exonerating these organisations, it is important to note that local NGOs and governments have often been at the forefront of promoting and implementing such displacement.

In countries like India for instance, local conservation strategies cannot always be philosophically differentiated from developed nation biocentric conservation strategies. India was not a clean slate on which the modern PA system was imposed. Hunting preserves that kept local communities out were a common feature of princely states; exclusionary strategies were also very much a part of British imperial policies there. Having said that, however, a significant membership of the conservation community is derived from upper-caste, upper middle-class urban backgrounds with historical sympathies to the exclusionary causes that have easily segued into contemporary sympathies for 'radical American environmentalism' (Guha 1989) and 'Edenic' philosophies (Robbins and Moore 2012). In the current conservation scenario, the cultural sacrosanctity conferred on PAs—as in the in the case of America, where PAs are designed to be inviolate—served as an easily adoptable role model; it remains an agenda that has been achieved through a combination of elite lobbying and top-down state initiatives. Decentralised conservation is logically inconsistent with inviolateness.

This quest for inviolateness, especially of PAs, came to a boil in the last decade starting in 2005 when the Schedule Tribe and Other Forest Dweller (Recognition of Forest Rights) Act (FRA) was drafted as bill. The tribal bill coincided with a crisis of missing tigers from a tiger reserve, causing two lobbies, one on behalf of tribal communities and another on behalf of the tiger, to come into conflict with each other. The conservationist resistance, gaining as it did contextual impetus from the crisis, is rhetorically designated here as the 'Tiger Lobby'. The drafting of the provisions of the FRA when it was a bill, when it became an Act, when the Act's rules were drafted, and when the rules were finally notified, all reflect the conflict, with each lobby at some point in the legislative process prevailing over the other by inserting favourable provisions or deleting unfavourable ones.

Nonetheless, the various delays can principally be attributed to the resistance of the Tiger Lobby, which was guided by a definite biocentric developed world ideology. Spatially, PAs were sought to be kept out of the FRA's ambit because the Forestry Rights Act seeks to vest land and community forest rights with an upper limit of 2.5 hectares per individual or family. Communities also can gain collective rights over community forest resources they have depended on and managed.

Many actors in the broad constituency, including scientists, conservation practitioners and policymakers protested the FRA and resisted its drafting and legislation. The Ministry for Environment and Forests (MoEF) in a centralised disposition, coupled with a sense of bureaucratic insecurity, argued that the FRA amounted to the removal of legal cover for forests that would result in massive deforestation. Several conservation organisations toed this decentralisation–deforestation line. For instance, the Bombay Natural History Society (BNHS) in June 2005 called for a discussion on the bill that it claimed 'would sound the death knell of our forests'.

Numerous apocalyptic future scenarios were put forward as the potential consequences of the FRA. This alarmist rhetoric predicted several potential outcomes: 60% of India's forests would be lost; tribal families would have unrestricted access to forest produce; existing wildlife and forest laws will be superseded; migrants will buy, lease or simply lure communities into parting with their lands; a market onslaught will ensue and ventures ranging from mining to hospitality will abound; the timber mafia will benefit from the spoils of deforestation; marijuana cultivation will thrive (Ramnath 2008).

Even though there has been clear misuse of the law, 10 years on, these predictions have not materialised. Several conservationists now support the law while pointing out its weaknesses and failures. Left to themselves, the Tribal Lobby, operating from a position of strength—gained from the fact that such a legislation was politically sanctioned in the first instance by a ruling majority coalition—would have seen the legislation through much quicker. Nonetheless, conflict still persists, with another legal battle wherein some groups are attempting to get the FRA struck down as unconstitutional. This push for the 'pristine' came not only from a Western notion of conservation but also from within.

Conclusion

Conservation can work at different scales, from the international to the national to the local. The impact of conservation philosophy and action can interact in complex ways across these scales. Without doubt, we agree with Rodriguez et al. (2007) that 'leadership in conservation has to be decentralized and integrated into local conditions'. However, the contrast with the 'local', on the other hand, is not as stark as one might believe. The local communities' knowledge of conservation politics and their ability to use it to their advantage is rarely considered.

As Redford states in Chap. 9, people use conservation in many ways, often to serve their personal interests. As our case studies illustrate, there is great variation and complexity in how and why conservation takes place and the myriad roles and approaches of the organisations involved are often at odds with the prevailing narratives of conservation.

The sectoral approach espoused by Brockington et al. in Chap. 3 could be one way to avoid falling into the common trap of idealising small groups and NGOs and villainising BINGOs as the source of all protectionist ideas around conservation.

References

Chapin, M. 2004. A Challenge to Conservationists. *World Watch* 17 (6): 17–31.

Dowie, M. 2005. Conservation Refugees: When Protecting Nature Means Kicking People Out. *Orion*, November–December: 16–27.

Guha, R. 1989. Radical American Environmentalism and Wilderness Preservation: A Third World Critique. *Environmental Ethics* 11: 71–83.

Ramnath, M. 2008. Surviving the Forest Rights Act: Between Scylla and Charybdis. *Economic and Political Weekly* 43 (9): 37–42.

Robbins, P., and S. A. Moore. 2012. Ecological Anxiety Disorder: Diagnosing the Politics of the Anthropocene. *Cultural Geographies* 20 (1): 3–19.

Rodriguez, J.P., A.B. Taber, P. Dasszk, R. Sukumar, C. Valladares-Padua, S. Padua, L.F. Aguirre, R.A. Medellín, M. Acosta, A.A. Aguirre, C. Bonacic, P. Bordino, J. Bruschini, D. Buchori, S. González, T. Mathew, M. Méndez, L. Mugica, L.F. Pacheco, A.P. Dobson, and M. Pearl. 2007. Globalisation of Conservation: A View from the South. *Science* 317: 755–756.

Shanker, K. 2015. *From Soup to Superstar: The Story of Sea Turtle Conservation Along the Indian Coast*. New Delhi, India: HarperCollins.

Shanker, K. and R. Kutty. 2005. Sailing the Flagship Fantastic: Different Approaches to Sea Turtle Conservation in India. *Maritime Studies* 3 (2) and 4 (1): 213–240.

Kartik Shanker works for Ashoka Trust for Research in Ecology and the Environment (ATREE), Dakshin Foundation and IIS.

Siddhartha Krishnan works for Ashoka Trust for Research in Ecology and the Environment (ATREE).

Marianne Manuel works for Dakshin Foundation.

15. Ed Tongson: The Lessons of Hands of Experience

What are some of the lessons from hands-on project experience? Let me illustrate some thoughts about conservation NGOs and the role of social sciences through one specific case. I have worked 18 years in the field of conservation with WWF Philippines.

This concerns our engagement with the Tubbataha Reefs located within the Coral Triangle at the centre of the Sulu Sea. The Tubbataha Reefs are the largest coral atoll formations in the Philippines covering an area of 100 km^2. The nearest land mass of size is mainland Palawan, with the capital of Puerto Princesa City lying 150 km northwest of Tubbataha. The islands of Cagayancillo are the nearest human settlement to Tubbataha and are home to the traditional users of Tubbataha's resources.

At the start of the 1980s, Cagayancillo fishers started to perceive the pressure of overfishing in their immediate surroundings. Threats came from increased commercial fishing and seaweed farming. In 1989, the near pristine condition of the reefs deteriorated because of illegal fishing, including use of explosives, indiscriminate dropping of anchors and unscrupulous collection of wildlife. Fishers from China and Taiwan also encroach into the Sulu Sea to catch turtles, live fish and sharks using destructive gear. The reefs also began attracting the attention of researchers.

Like any conservation project, the case of Tubbataha Reefs started with conflicts between what we broadly categorised as 'preservationists' and 'extractionists'. The WWF, a conservation NGO, was identified with the

P.B. Larsen, D. Brockington (eds.), *The Anthropology of Conservation NGOs*, Palgrave Studies in Anthropology of Sustainability, DOI 10.1007/978-3-319-60579-1

preservationists. It allied itself with the dive industry, which had the means to mobilise resources to protect Tubbataha, and with the academe who were into research. At the opposite end of the spectrum were the big and small-scale fishers, the latter being supported by Palawan-based NGOs.

At the start of the Tubbataha project, this writer remembers we had a better agreement in our approach with the preservationist NGOs (Saguda, Sulu Fund, dive industry) than with extractionists that included the Palawan-based NGOs (PNNI, ELAC) that favoured maintaining rights of small fishers to fish in Tubbataha. Getting the cooperation of the preservationists to declare Tubbataha as a no-take marine protected area (MPA) was not difficult. The late 1990s was the lowest point in the history of Tubbataha as access to resources was a free-for-all. There were Chinese turtle poachers, seaweed farming, blast fishers, bird hunters, commercial fishers, shark hunters and all types of destructive fishing. The preservationists witnessed the destruction of Tubbataha and would welcome any enforcement action from any group. But enforcement in Tubbataha will be logistically expensive if stakeholders were not involved, especially fishers in Cagayancillo, who will lose out in case Tubbataha becomes a no-take MPA.

The conservation agenda always will have short-term consequences to livelihoods for small-scale fishers. Although they understand the long-term benefits of conservation, artisanal fishers cannot wait and are motivated by short-term gain. This is the reason why the entry of big conservation groups is unfavourably received. Building trust and goodwill among Cagayancillo fishers therefore was an important part of the strategy in Tubbataha. Our grants were channelled to improving seaweed-based livelihoods and creating local sanctuaries to allow fish stocks to grow and reproduce so that local catches could improve. With trust and goodwill, it was not difficult for fishers to voluntarily waive their rights to fish in Tubbataha. The Palawan NGOs, who were more concerned about fishers' rights being curtailed by impositions from more powerful groups, deferred to the fishers' wishes if they perceived the negotiation process to be fair and transparent.

Enforcing a no-take zone in the middle of nowhere can be logistically expensive. This is compounded by the need to support community projects in Cagaynancillo to dissuade them from going back to fishing in Tubbataha. Sustainable financing through user fees was the best option. The WWF brought in resources from grants from US donors and secured commitments from dive operators and diving groups to support a user fee system. This was to defray the cost of managing the park and sustaining future needs. A compensatory scheme was proposed to the small-scale fishers in exchange for waiving their rights to fish in Tubbataha. The arrangement involved support for community projects.

I agree with the observation that conservation NGOs have much to learn from social scientists. Many conservation projects fail because local stakeholders share a disproportionate burden of the cost arising from a no-take zone compared to benefits accruing to global and national stakeholders and more powerful groups. In our case, we benefitted from the advice of a sociologist who early on introduced us to the tools and techniques (Technology of Participation, stakeholder analysis, participatory research, equitable sharing of costs and benefits) that became part of our conservation toolkit.

The stakeholder analysis is performed at the beginning to map each stakeholder group as to influence and/or power and interest that would point to potential areas of conflict. The technology of participation consists of structured facilitation exercises where stakeholder groups think, talk and work together to manage conflicts leading to equitable sharing of costs and benefits. Participatory research is an approach that emphasises collective enquiry, experimentation and learning grounded on experience and reflection. They are widely used in the portfolio of projects that we handle throughout the country. I would add that conservation NGOs also can benefit from a strong human resources development and training program (as found in most corporations) where new hires in middle management are trained on the standard set of best practises and use of the conservation toolkit. This will prepare them to implement field projects on their own, especially on how to deal with stakeholders.

Ed Tongson is Vice-President for Sustainable Technologies for WWF Philippines.

16. David Wilkie: Looking Over Fences will not Promote Engagement

Public, third-party critiques of not-for-profit civil society organisations are vital to promote their effective governance. These critiques also provide an additional valuable layer of transparency and accountability to the NGOs for their partners, supporters and staff members.

The set of chapters selected by Peter Bille Larsen and Dan Brockington are a critique of conservation NGOs that is almost as broad in its focus as is the social science discipline of its authors. All are those authored by social scientists and the single one by a conservation scientist raise important and valid issues that the conservation community and its policymakers and practitioners should attend to. Sadly, as written, most will not.

Rather than attempt to respond to all issues in all chapters, I would like to focus on three. The first concerns the authors' use of impolitic analogies and language that risk undermining social scientists' contributions to conservation governance and practise; the second relates to the purported power of the BINGOs, and the third refers to the assertion that the conservation community has adopted an exclusive, neoliberal, market-fixated philosophy.

First, for these chapters to influence the philosophies and practises of conservation NGOs their messages must not only resonate with the authors' social science peers, but they also must be taken seriously by

P.B. Larsen, D. Brockington (eds.), *The Anthropology of Conservation NGOs*, Palgrave Studies in Anthropology of Sustainability, DOI 10.1007/978-3-319-60579-1

conservation practitioners that are keen to strengthen the governance of their organisations. To do the latter, they must befriend rather than alienate. An edited book titled *The Anthropology of Conservation* would suppose a respect for other cultures. Yet titling and structuring a chapter around conservation practitioners as analogous to Clint Eastwood in two violent, vigilante movies does nothing other than cause the conservation practitioner reader to frown in amused disbelief. In a misguided attempt to be cute or clever, this pointless analogy only serves to undermine the authors' valid concerns.

Similarly, equating conservation–private sector engagement with a Japanese martial art that uses an opponent's own power against them, is neither an accurate nor useful analogy. In addition, saying that conservation NGOs are 'enslaved to their donors' priorities' (p. 124) is just startlingly crass because it suggests that conservation organisations have no agency, and worse are tone-deaf to those who were once and are currently truly enslaved.

Second, a common thread woven through this book's chapters is the power, largely assumed illegitimate, of the BINGOs to dictate uses of lands and waters around the globe in the name of conservation. It is true that they mobilise and incur expenses of approximately USD1.5 billion each year for conservation actions. However, as Larsen noted, by citing Waldron (2013), total spending on conservation exceeds USD19.8 billion annually. As the BINGOs are spending less than 7% of global investment on conservation it would be remarkable if they wielded the power attributed to them in these chapters in terms of their purported outsized influence on all conservation actions and land use decisions globally.

Although Overseas Development Assistance (ODA) in 2015 exceeded USD131 billion, it is difficult to believe that the BINGOs are exerting as much influence on the use of the planet's lands and waters as are the development NGOs and governments who are spending ODA funds each year. Yes, conservation BINGOs are important influencers of how the planet is used, but they are not the most powerful. Social scientists are well positioned to look critically at all the major players that influence the use of our planet and that respect or prejudice the rights of indigenous and local people. Focusing on the small fish in the pond seems like purposeful underachievement.

Third, reference to neoliberal conservation and an interest in markets by the conservation community is an exaggerated and incomplete story at best. Contrary to several authors' contentions, the conservation community has not abandoned its conservation priorities to embrace market solutions. Rather, I would argue, the conservation community has added one other tool to its toolkit (i.e., engagement with the private sector). All conservation NGOs, regardless of what their websites may say, are staunch supporters of government-regulated PAs. Public and private protected areas across all six IUCN categories are all subject to de jure regulations and all preference nature conservation over all other land and resource uses. They require government regulatory and fiscal interventions and NGOs understand this.

In states without the means or interest to invest in PA network management, NGOs have and continue to seek ways to contribute to their protection and, most important, to build government capacity and interest to invest in the conservation of their protected area systems (i.e., NGOs are facilitating government interventionism rather than encouraging their withdrawal from the PA 'marketplace'). Nonetheless, conservation NGOs understand that PAs only cover a small percentage of the planet's lands and waters and alone cannot be expected to protect all biodiversity and natural resources that are intrinsically or tangibly valued by humanity.

Understandably then conservation NGOs were drawn into those spaces outside of protected areas where economic development was the primary land use objective. Engaging the private sector and attempting to avoid or mitigate their environmental impact is fraught for the conservation community (i.e., the reputational risk of 'sleeping with the enemy') and, given the economic and political power inequities, has only limited the probability of success. The conservation community, however, is unanimous that ignoring the influence markets have on resource use and hiding from purposeful engagement with commodity producers, manufacturers and consumers will result, without doubt, in the loss of biodiversity and the impoverishment of natural resource dependent indigenous, traditional and local people. As Mark Twain so sagely noted 'denial is not just a river in Egypt,' so not engaging with market-focused actors outside of PAs is not an option for the conservation community.

Finally, I must agree with Kent Redford that social science critiques based on looking over the conservation fence rather than talking and working directly with conservation practitioners in the field always is likely to result in misunderstanding and misinterpretation of what the critics perceive solely from afar.

David Wilkie is Executive Director, Conservation Measures and Communities at the Wildlife Conservation Society.

Index[1]

[1] Note: Page numbers followed by 'n' denote end notes.

P.B. Larsen, D. Brockington (eds.), *The Anthropology of Conservation NGOs*, Palgrave Studies in Anthropology of Sustainability, DOI 10.1007/978-3-319-60579-1

T

U

Zeitfracht Medien GmbH
Ferdinand-Jühlke-Straße 7
99095 Erfurt, Deutschland
produktsicherheit@kolibri360.de